FREEFALL

"FREEFALL

[by]
William Hoffer
AND
Marilyn Mona Hoffer. "

363.124

St. Martin's Press
New York

c 1989

Design by Susan Hood

Library of Congress Cataloging-in-Publication Data

Hoffer, William.
    Freefall : a true story / William and Marilyn Hoffer.
      p. cm.
  "A Thomas Dunne book."
  ISBN 0-312-02919-5
    1. Aeronautics—Canada—Accidents—1983.   I. Hoffer,
Marilyn. II. Title.
TL553.5.H64 1989
363.1'2416—dc19   89-30132
                                       CIP

First Edition
10 9 8 7 6 5 4 3 2 1

To Mert and Joe
and
Lucile and Frank

This is a true story.

1

The Cockpit

Captain Robert Owen Pearson, solidly built, his tanned face dominated by dancing eyes and an ever-present mischievous grin lurking beneath a clipped dark mustache, swallowed the last morsels of his dinner and sat back to relax. Air Canada Flight 143 was en route from Montreal and Ottawa to Edmonton on this late afternoon of Saturday, July 23, 1983. As he approached a point 41,000 feet over Red Lake, Ontario, Pearson was lulled into serenity by the familiar, steady roar of the thin atmosphere as it was parted, knifelike, by the nose of a 132-ton Boeing 767 forging through the sky at 469 nautical miles per hour (knots). Here in the cockpit the noise was far louder than back in the passenger section, and consequently more impressive. It was a constant reminder of the power that propelled this behemoth along in an environment alien to the natural powers of humanity.

Below was a soft, cottony cover of cumulonimbus clouds. Behind was the noise and bustle of Montreal. Ahead was Edmonton, Alberta, the gateway to the Canadian Rockies. Above was pure blue.

Pearson, a twenty-six-year veteran with Air Canada, a crown corporation owned and operated by his government, leaned forward slightly and flicked a button, switching the Air Traffic Control (ATC) channel on his radio to a cockpit speaker so that he could pull off his earphones and chat with the first officer. On a short flight, such as the leg he had flown earlier this day from Montreal to Ottawa, there was little time for chit-chat. But on this longer, four-hour flight from Ottawa to Edmonton, as the passengers were treated to dinner and a movie, Pearson could indulge in one of his favorite pastimes. He loved to talk. Spend five minutes with Bob Pearson and he would inevitably ingratiate himself with the glib, easy manner of his steady stream of conversation, punctuated by the colloquial "eh?" interjected at the end of nearly every sentence, that identified him as an English-speaking Canadian.

"That Ottawa leg is busy, though, eh?" Pearson said to First Officer Maurice Quintal, seated on his right.

Quintal, who had turned thirty-six two days earlier, was a soft-spoken French-Canadian whose impish appearance suggested a naïveté that masked the pathos of his personal life. He quickly agreed with his captain.

"It's good for us," he acknowledged.

"Sure it is. Yeah. Gets you sharp, eh?"

"Yeah."

Pearson and Quintal had flown together only a few times prior to this day, but they had developed a fledgling friendship. Pearson, active in the pilots' union, the Canadian Air Line Pilots' Association, had once helped Quintal with a difficult, job-related problem. He had, in fact, saved Quintal two weeks' pay.

"It's good for a guy who flies, uh, once every two weeks," Pearson continued, in reference to Quintal's position as a reserve first officer, flying only a handful of flights per month. Pearson regarded the aircraft in which he sat, and reflected with pride that he and Quintal were among the few pilots in the world qualified to fly this new beauty, the world's most modern airliner.

"Everything's straightforward once you learn it," Pearson commented. "It's not something you're likely to forget—all this stuff—either, eh?" To be sure, anyone who had experienced the wonders of flight in the cockpit of a spanking-new Boeing 767 was unlikely to forget it; there was a distinct resemblance to a video game emporium. With the introduction of the aircraft, Air Canada had announced the age of "digital avionics."

To the uninitiated, the 767 flight deck appeared complex and confusing, loaded with the expected array of dials, switches, pedals, circuit breakers, and twin control yokes. But the veteran saw something different, as exemplified by the military pilot who, upon his first glimpse of a 767 flight deck, commented, "They've simplified things a lot, haven't they?" That simplification was a direct result of the computer age. Video screens replaced dozens of old-style instruments and reduced eye strain, particularly on a long flight.

Among all Air Canada pilots, Pearson was known for his ability to capitalize on the new generation of electronic wizardry that was on display here. An entertainer by nature, the captain took advantage of every opportunity to bring co-workers, friends, and occasional passengers to the cockpit during flight. "You wanna see a light show?" he would ask rhetorically, grinning, flicking switches, coaxing an array of computer screens to life. The gregarious captain particularly enjoyed impressing children, whose eyes matched the colorful computer displays twinkle for twinkle. "How are you?" he would query, chucking a six-year-old under the chin, "Where are you going? How do you like flying? It's great, eh?" Words tumbled out of Captain Pearson.

If, on this day, Pearson was in the mood he described as "fat, dumb, and happy," it was only normal. Yet his contentment was heightened by a special opportunity offered on this particular trip. The day's flight would be followed by a twelve-hour layover in Edmonton. Then, tomorrow, he would welcome his in-laws on board for the return trip. Although their daughter Carol had been married to this commercial pilot for twenty-five years, Percy and Gwen Griffith had remained reluctant to fly, preferring to use the pass Percy had earned from his career with Canadian Pacific Rail. As a result, Gwen had flown only once and this would be Percy's maiden voyage. Their disinclination was an attitude that Pearson found difficult to comprehend, and he had vowed to himself to do something about it. He would see to it that they enjoyed themselves, inviting them up to the cockpit to view the ease with which man's understanding of physics, coupled with basic engineering skills, and now further enhanced by electronic technology, conquered the sky. It was fortuitous that the weather had cooperated so beautifully.

As Flight 143 roared through the sky, the 767's bank of computers not only reported the smooth progress of the uneventful journey to Pearson and Quintal, but also to a vast array of computers on the ground. Sometimes, it almost seemed that a 767 pilot was unnecessary.

This aircraft, Number 604, was one of four 767s that Air Canada had purchased for $40 million apiece; seven more were on order. Only three months old, she was an awesome machine with a wingspan wide enough to cover half the length of a football field (even a Canadian football field, ten yards longer than in the U.S.), and capable of holding nearly 20,000 U.S. gallons (more than 50,000 kilograms) of Type A-1 kerosene jet fuel. On the ground, the tip of her tail would rise above a five-story building.

But the ground was not her natural environment. Two Pratt & Whitney fanjet engines, swallowing 4,000 kilograms of kerosene an hour, produced a combined thrust of 96,000 pounds, enough power to propel the airliner along the airways at 470 knots (about Mach 0.8) over a range of about 10,000 kilometers (the range that Air Canada pilots used for their calculations) or about 6,000 statute miles (the range that the pilots used in their heads). At maximum thrust, the tips of the eight-foot fan blades actually revolved at a rate faster than the speed of sound.

To Pearson and the handful of other pilots qualified to fly her, the 767 was a joy to handle. The cruising altitude of either 39,000 feet or 41,000 feet or even 43,000 feet was considerably higher than that of older generations of jetliners. The 767 could fly in the smoother altitudes of the lower stratosphere, above the strongest jet stream winds. Often, Pearson found himself flying calmly over weather that had other pilots scurrying for detours.

Newness has its tradeoffs. Despite his unabashed love of the aircraft, Pearson found the 767 a bit of an enigma. This was a man who could fly just about anything with wings, and he had long ago learned to appreciate and trust the science of mechanics, but, despite his penchant for showing it off, the 767's newfangled electronic technology left him feeling vaguely uneasy. A pilot's bread-and-butter, his very existence, is *control*. Computers usurped a measure of that control, hiding vital information in their electronic innards that Pearson preferred to have in his own head.

By 1983, although computer technology had progressed exponentially, its mysteries were still controlled by engineers and programmers. The nontechnical citizen was still unfamiliar with—and somewhat intimidated by—the computer. Pearson once muttered to his wife Carol, "We'll probably be the last house in Beaconsfield to have a computer."

He knew that these gentle misgivings were reactionary; like everyone else, he had to adjust to life in a technological age. He had been in love with the sky for thirty-one years, and from his viewpoint high above the earth, he had seen incredible change.

The love affair began in 1952 when—not yet sixteen—Pearson spent the summer working in the northern bush of the province of Quebec. In those days Indians in Quebec were not permitted to buy or consume any type of alcoholic beverage. Pearson could not countenance such discrimination; he empathized with the Indians' thirsty plight. On weekends, homesick for some English conversation, he would make the two-hour drive to Baie Comeau on the north shore of the St. Lawrence River, a frontier town where a teenager could purchase beer with no questions asked. He would return to the bush on Sunday evening toting several extra quarts for his comrades. Not surprisingly, the Indians quickly accepted him into their fold.

There came a magic day that summer when, along with thirty or forty locals, some French Canadian, some Indian, the young Pearson was drafted to help fight a forest fire. Most of the firefighters were loaded into trucks for a spine-jolting ride through the rugged bush, whereupon they would still face a long hike to the site of the fire, but a few had an alternate mode of transportation available. One of the Indians asked the youngster if he had ever been in an airplane.

"Nope," Pearson replied.

"Come on with me," the Indian responded. Pearson crawled into the back of a De Havilland Beaver float plane along with six Indians. The single-engine, high-winged workhorse of a plane was outfitted to carry forty-five-gallon fuel drums, not people. Consequently, there were no seats or seatbelts in the back. All that stood between this

6

rag-tag group of firefighters and the great outdoors was a makeshift door latch, similar to one on a kitchen cupboard.

Approaching the fire zone, the adventuresome bush pilot decided to buzz the fire tower. He tilted the plane, first on one wing tip, then the other. Pearson and his terrified co-workers ricocheted back and forth, like so many steel balls in a pinball machine. The pilot brought the plane in low over the forest fire. Pockets of smoky air buffeted them. All about him, Indians screamed in terror, but Pearson thought: This is the greatest thing I've ever done.

He was hooked.

Back in his native Montreal two years later, Pearson took a job as an electronics technician at McGill University. "I hated it," he said later. "I hated being inside. I was a clock watcher." Nevertheless, Pearson put in his hours on the job so that he could climb into a cockpit; he soon found, however, that his paycheck was insufficient to fund the fourteen-dollar-per-hour flying lessons. He borrowed money from his father. He sold his car and anything else that was marketable to come up with the necessary cash.

Within a year, he had qualified for his commercial license and passed the required examinations. To be hired by a commercial airline he needed 200 hours of flight time, and his logbook certified that he had 200 hours and six minutes. All he had to do now was find someone to hire him.

In May 1957 Pearson learned that Trans Canada Airlines, the government-owned passenger carrier, was hiring pilots. The only hitch was that he was still waiting for that necessary piece of paper—his commercial license—to arrive in the mail. Unable to wait, he called Trans Canada and made an appointment for an interview, praying that his license would arrive on time. Each day he checked the post office box anxiously, to no avail. Finally, on the appointed day, as he walked from his parents' home to the railway station he stopped at the post office, peered through the glass door of the box, and spied the brown envelope waiting for him. Muttering thanks to the deity of flight, he slipped the precious piece of paper into his pocket, continued on to the offices of Trans Canada, and was hired.

Five months of training followed before Pearson qualified as a first officer on the DC-3, the twin-engine, propellor-driven craft that had long been the workhorse of the industry, but was being phased out by turbine-powered aircraft such as the Vickers Viscount and the Lockheed Electra.

Over the years, as Trans Canada grew, changing its name to Air Canada, Pearson's career soared higher. Partly for excitement and partly in pursuit of promotion, he volunteered for training on each new aircraft type as it came along. Not all commercial pilots are so venturesome—many prefer to wait until the wrinkles have been ironed out of a new airplane; others remain comfortable on one type of craft. But Pearson moved from cockpit to cockpit, relishing the diversity. He progressed steadily from the first officer's seat of the DC-3 and the Vickers Viscount, to the Douglas DC-8 and DC-9, and the Boeing 727, as they became available.

In 1967 Pearson was promoted from senior first officer to junior captain, an upward move that, ironically, relegated him to a lower priority position. He returned to the Viscount, this time as a captain, and gradually worked his way back up to the jetliners, logging considerable time on the twin-engine DC-9 and the triple-engine 727.

On February 21, 1983, Pearson began a two-and-a-half-week ground school training course, taught by Air Canada in Montreal, to fly the new Boeing 767. He sat by himself in front of a teaching machine that allowed him to progress at his own pace, learning the intricacies of the 767 Flight Manual. It was exciting, for Pearson knew that the manual was necessarily incomplete. He and others operating the new aircraft would participate in revising it, learning from experience.

8

Over the years, Air Canada's flight training had evolved to keep pace with advancing technology. In the old days of simpler aircraft, training was governed by a "nuts and bolts" philosophy. Not only did the pilot learn how to operate every component of the aircraft, he also learned the mechanical and physical principles behind each one. On some types of aircraft, the pilot had to be able to draw diagrams of the electrical, pneumatic, and hydraulic systems. The advent of sophisticated jet aircraft changed all that. Gradually the nuts and bolts method of training gave way to a "need to know" philosophy. The pilot was required to understand those functions over which he maintained control, but he was neither expected nor encouraged to concern himself with the myriad systems that were now the province of others. The 767 took this concept to an advanced stage. Pilots would fly it, mechanics would fix it, and neither group was required to comprehend the duties of the other.

The fueling system provided a prime example of this evolutionary growth in training. On a 727 or a DC-8, for example, pilots were taught how the various fuel pumps were operated, where the fuel lines were located, and how overflow or venting pipes were installed. In contrast, 767 pilots merely were informed that a computer known as the fuel quantity processor supplied the fuel gauges with an accurate reading of the quantity of kerosene in the tanks. They had no need to know how it worked.

There was one particularly troublesome and somewhat confusing aspect to the training. The aircraft's specifications were designated by a mix of metric and imperial numbers. Fuel was calculated in kilograms. Thrust was designated in pounds. "We were just learning numbers," Pearson commented. "They could have been bananas."

This was also confusing to the mechanics, who needed two sets of tools—both metric and imperial—to work on the aircraft.

9

Ground school was followed by two weeks of training in a flight simulator in Toronto and a day flying "circuits"—takeoffs and landings—at the Toronto International Airport. This was a heady experience, for the 767 boasted a far higher power-to-weight ratio than any other commercial airliner. The takeoff run was short and oh-so-sweet. The rate of climb, compared to other jetliners, was breathtaking.

After Pearson's single day in the 767 cockpit, an official of Canada's Ministry of Transport checked him out and, on April 3, 1983, licensed him as a 767 captain.

Two and one half months later, by July 23, Pearson had logged more than 15,000 hours of professional flying time in a wide variety of airliners. Fewer than 200 of those hours were spent in the cockpit of a 767, but, because the plane was so new, he qualified as an expert.

At the age of forty-eight, nearly a quarter of a century after his first raucous flight over the Canadian bush, Pearson's unique combination of skill and moxie had established him in a secure routine. His children, Donna, Glen, and Laurie, were nearly grown and on their own. His wife, Carol, had found her niche by returning to John Abbott College in Ste. Anne Bellevue, studying art history and designing pottery, but, as she always had, she made her husband her first priority. When he was away, she busied herself with books and pottery, working at a studio in the nearby Montreal suburb of Baie d'Urfe. When he was home, she busied herself with him. She tended not to think of a day as a Monday, Tuesday, Wednesday, or whatever; rather, the week was divided into "days when Bob is here" and "days when Bob is not here." Carol once said, "I sometimes think I've got everything so good, some day there will be a big bang."

The Pearsons' nest was nearly empty (except for the pet rabbit known as Sherman McCloud, or B.R.—"bunny rabbit"—for short) and they gloried in their freedom. On a fifty-

10

acre farm near Dalkeith, Ontario, land they had purchased in 1959, Pearson had just begun to set into place the first few logs for a home he planned to build from trees he had planted more than two decades earlier. He and Carol looked forward to a comfortable retirement in the distant future.

Although he relished the flexibility of the pilot's schedule, which allowed for many more days on the ground than in the air, Pearson was not one to sit still. One passion of his life was the Montreal Air Canada Pilot's hockey team. Only recently he had added skiing to his repertoire of skills. More often, he was inclined to take a busman's holiday, climbing into the cockpit of a private plane. For years he had combined his commercial flying with work as a glider instructor and tow-plane pilot. He enjoyed the feeling that he was master over the aircraft, that he could make it conform to his will.

As the autopilot kept the aircraft smoothly on course, Pearson and First Officer Quintal listened with amusement to other pilots in older-generation aircraft requesting permission from Winnipeg Air Traffic Control to alter their altitudes. Although the weather report had predicted mostly clear conditions over central Canada, there was a bit of rough air below, a front of cumulonimbus clouds providing the kind of unpredicted turbulence that underscores the advice to keep one's seatbelt fastened during flight. Pilots of older aircraft jockeyed for position.

It was 7:57 P.M. in western Ontario far below, but 0057 in the more useful airborne scale of Greenwich Mean Time (GMT), when Pearson radioed the Air Canada office at Winnipeg International Airport. "Winnipeg, 143," he said. "Would you go ahead the . . . the latest Edmonton weather, please?"

A controller's voice filled the cockpit: "Midnight Edmonton, 3,500 scattered, 30,000 thin broken, fifteen miles, temperature twenty-four, dewpoint fifteen, 170 degrees at seven, and the altimeter 999."

"Thank you kindly, Winnipeg," Pearson drawled. He turned to Quintal, who was jotting notes. "Get that?"

"Yeah." It was close to perfect. Fifty-seven minutes earlier, at midnight GMT, or 5 P.M. in their destination city of Edmonton, the airport was lightly covered with scattered clouds at 3,500 feet, topped by a broken cloud cover at 30,000 feet. Visibility was fifteen miles. The temperature was a pleasant twenty-four degrees Celsius, or seventy-five degrees Fahrenheit. Winds were seven miles per hour, emanating from 170 degrees, or south-southeast.

Quintal told Pearson that he was going to update the passengers on the easy progress of the flight.

"Okay, yeah," Pearson said. "I'll watch the boat."

Pearson switched off the cockpit speaker so that ATC transmissions would not interfere with Quintal's announcement to the passenger cabin. He pulled his earphones into place, grinned, and quipped, "I'm going to sit here and watch the trout swimming in the lakes."

Quintal activated the public address system. His calm, professional voice, spiced with a French-Canadian accent, broke into the reverie of the passengers as they finished their dinners. "Good evening, ladies and gentlemen, this is your first officer," Quintal said. "We're presently coming up over Red Lake, presently 800 miles from Edmonton, cruising at 41,000 feet. The temperature in Edmonton is . . . beautiful day, clear, temperature of twenty-four degrees Celsius. Thank you." He repeated the announcement in his native French.

Pearson and Quintal both turned as a visitor entered the cockpit. Dark and stocky, forty-one-year-old Rick Dion was a private pilot who had dreamed of occupying the captain's seat in a cockpit such as this, but instead had risen to the higher echelons within Air Canada's maintenance system.

Two days earlier Dion had finished a five-month assignment that was a testimony to his expertise. With the growth of the computer age, Air Canada's maintenance functions gradually became centralized. In the old days each major airport had maintained an office known as Maintenance Central, which was the ultimate authority on the mechanical, electrical, hydraulic, and other operating systems for aircraft being serviced at its location. Computers, with their capacity to share volumes of information at the speed of light, had enabled the airline to coordinate its maintenance functions into a single office, located at Montreal's Dorval Airport and known by the stronger term Maintenance Control. There, Dion worked as a troubleshooter, answering queries from mechanics all over the world, which were relayed by telephone, teletype, computer, and an open conference-call circuit. Dion's assignment there indicated the trust and respect of management.

Today he was embarking with his wife Pearl and their three-year-old son Chris on a quick trip to Vermillion, Alberta, a small town in the rolling hills east of Edmonton, to visit Pearl's eighty-one-year-old father.

Before boarding the flight in Montreal, Dion had spotted Pearson and said hello. They were old acquaintances, having met at Cooper Aviation, a private airport where Pearson was a partner in a franchise that sold ultralight aircraft. Pearson had invited Dion to the cockpit after dinner; they could talk shop. The mechanic accepted the invitation immediately, for he was unfamiliar with the new 767, unqualified to work on it, but excited and interested to view it in action.

Just as Dion arrived, Quintal decided to stretch his legs for a moment. "So you're going to go back and do a little jogging?" Pearson asked.

"Right," Quintal replied.

Pearson slipped into an exaggerated Italian accent, "You wanna me to get everybody to sit-a down, clear-a the aisles?"

"Yeah, go ahead, do that." Grinning, Quintal left.

Pearson fiddled with his newfangled digital watch, attempting to calibrate it with the aircraft's liquid crystal display. "I have to reset my goddamn watch so seldom that I keep forgetting how to do it," he muttered to Dion. "It's harder than flying this airplane."

Dion's eyes quickly took in the features of the new cockpit. On the ground in Ottawa prior to the second takeoff of the day, he had participated in discussions concerning two seemingly minor malfunctions, the kind that are routine in the life of a commercial pilot.

One of those problems was the failure of the fuel quantity processor, and now that he had a chance to view the cockpit during flight, Dion's trained mechanic's eye wanted to see how the problem affected operations. He spotted what he believed were the fuel gauges. The gauge for the right-hand tank was blank, but what he thought was the left tank gauge read "six-point-something." That was consistent with an entry in the aircraft's logbook, in its holder between the pilots' seats, that indicated one of the fuel processor channels to be "inop."

In fact, all of the fuel gauges were inoperative and, had Dion realized this, he might have questioned the airworthiness of this particular 767. As a troubleshooter for Maintenance Control, Dion knew that all other types of aircraft would be grounded in this condition, for there would be no way to monitor fuel supply during flight. But this was a new 767, with its computerized, intentionally redundant systems. Dion made the natural assumption that it carried enough special equipment to compensate for blank fuel gauges. In any event, he would have had to check the Minimum Equipment List (MEL), which he regarded as the bible. To him, the MEL was the final authority dictating the conditions whereby an aircraft could be—or could not be—cleared to fly.

If Dion had had any reason to consult the appropriate passage in the 767's MEL, he would have discovered to his

14

befuddlement that, with all the fuel gauges blank, the very aircraft that now carried him and his family over western Ontario should, in fact, have been grounded in Montreal.

But the mechanic did not know this, nor did the captain and first officer. In fact, what Dion believed to be the left tank fuel gauge was merely a thermometer measuring the temperature of the fuel in the tanks, reporting it on the Celsius scale, still somewhat unfamiliar even though four years had passed since Canada had "gone metric" in 1979.

Dion was one of many Canadians who grumbled that "metrics was jammed down our throats." Four years after metrication began, the mandated change was still a controversial issue. A noted case was even now pending in a Toronto court, wherein a butcher was being prosecuted for selling meat by the pound, rather than the kilogram. In Alberta, vandals regularly defaced metric road signs. For his part, Dion still found it disconcerting to stop at a service station, fill the car's tank with liters rather than imperial gallons, and drive off at a speed limit of so many kilometers per hour. "I still have problems with meters and centimeters and millimeters," he would admit.

Air Canada had ordered its four new 767s built to metric specifications, partly as a result of pressure from the government, and this had brought predictions of trouble from the loyal opposition. Neil Fraser, a Conservative party leader who had campaigned hard against metrication, argued, "If Air Canada wants to experiment with metric fueling, I suggest it does so with the private jet used by the federal cabinet."

But at this moment the metric fuel gauges were not Dion's primary concern, for he was confident that adequate safeguards were in place. In fact, on separate video monitors in front of the captain and first officer's seat, the flight management computer had taken over for the blank fuel gauges, depicting a succinct summary of the fuel load.

What Dion wanted to discuss was a second operational conundrum, the type of snag that commonly occurs as the bugs are worked out of a new airplane. During the first leg of the flight, the right engine bleed valve warning light had illuminated, and Dion and Pearson now pursued the subject further. On each of the two engines a bleed valve regulated the amount of air diverted to the cabin air-conditioning system. If the bleed valve did not work properly cabin temperature might soar. Worse, the airframe deicing system might lose effectiveness.

When the warning light illuminated, Pearson and Quintal had quickly followed the dictates of their training, consulting the operating manuals that detailed the appropriate response. They first checked the Quick Reference Handbook for a concise explanation of emergency procedures. It informed them that as long as only one bleed valve warning light was on, they were to take no action unless there was a need to activate the deicing system. But later, on the ground in Ottawa, they had time to consult the more detailed MEL, which declared that they should have activated the Auxiliary Power Unit (APU).

As it happened, the problem had cleared up by itself. The bleed valve was now functioning normally, but the pilot and mechanic were concerned over the conflict between the Quick Reference Handbook and the MEL.

"There's a discrepancy," Dion said.

"A big one," Pearson acknowledged.

"Well, that's something to bring up. This is a new airplane. You're gonna run across that, you know. So that's maybe a good point to bring up."

Pearson agreed, and made a mental note to submit a report to his superiors. Such a discrepancy was not unusual in the operating manuals for a new type of aircraft. Pilots and mechanics had to work together to iron out the proper procedures. Pearson needed a definitive answer. Should

16

he or should he not activate the APU, if and when the bleed valve warning light illuminated? His primary interest was that the Quick Reference Handbook reflect the proper course of action, because that was the document that any pilot would consult immediately in the event of an emergency. "That's all we're concerned about," he said. "When something happens in flight, wham! We got a . . . we got a drill. That is the one and only drill we do."

"Right," Dion agreed.

Pearson gestured at the MEL, much thicker and more ponderous than the Quick Reference Handbook. The MEL would be consulted if a problem occurred on the ground prior to takeoff, but it was far too voluminous to be of use during an in-flight emergency. "This fuckin' thing," he said, "we may or may not take it out of the case."

The pilot and the mechanic continued the discussion amiably, two professionals working together to improve the system. Dion pointed out, "Listen, these MELs change as time goes on and, uh, you know I'm sure, like, some of the older airplanes we have, the original MEL is far different than what we've got now." He noted that the lessons of experience caused Air Canada to continue to make changes in the MELs for the Lockheed L-1011 and the Boeing 747, both of which had been flying for many years, and the 767's MEL was only beginning its evolutionary growth.

In the midst of this discussion Quintal returned from his break. Seeing that Pearson's attention was occupied, and that he had both the Quick Reference Handbook and the MEL open on his lap, he asked if Pearson had jotted down the required notes as they passed abeam of their latest checkpoint, Red Lake.

"No, I didn't get that number," Pearson said.

So Quintal recorded the information on the flight plan at this, the third of four checkpoints between Ottawa and Edmonton. Flight 143 was seven minutes ahead of schedule,

17

which was satisfying, since it had left Montreal twenty-eight minutes late due to difficulty with the fuel gauges. Thanks to a shorter than usual stopover in Ottawa, it had left there only eight minutes late. A light load of passengers on this lazy Saturday afternoon had enabled Pearson and Quintal to request and receive clearance to cruise at the flight's maximum altitude of 41,000 feet instead of the planned 39,000 feet, and this had allowed them to cheat the clock further.

These early days on a new aircraft had been frustrating. On-the-job training seemed the norm for a 767 flight. Malfunctions and warnings of perceived malfunctions were the rule, rather than the exception, and they frequently proved to be problems within the computer systems, rather than the mechanical systems.

For once, Pearson thought, everything is working right.

As required at each checkpoint, Quintal now compared the present fuel load with the minimum dictated by the flight plan. Here over Red Lake the minimum fuel requirement for reaching Edmonton safely was 8,500 kilograms of Type A-1 kerosene jet fuel. Quintal glanced at his video terminal of the flight management computer, acting as a backup for the blank fuel gauges. The computer indicated that the wing tanks of the 767 carried 11,800 kilograms of fuel, a comfortable 3,300 kilograms above the minimum at this point.

Just as Quintal finished recording this information on the flight plan, at 0109 GMT, a sharp warning buzzer emitted four short beeps in the space of a single second that commanded the attention of the three men in the cockpit.

Dion thought: Something's wrong!

"What would that be?" Quintal wondered aloud.

Pearson, never at a loss for words, reacted more vehemently. "Holy fuck!" he exclaimed.

18

2

The Cabin

One hundred fifty feet back from the cockpit of Flight 143, at the very rear of the aircraft, flight attendant Susan Journeaux Jewett began to work her way forward from the galley. She pushed a cart through the right aisle of the aircraft, clearing away littered dinner trays and taking orders for after-dinner drinks. She glanced at her watch. Although her body was 41,000 feet over central Canada, her heart was back home in Montreal. She wondered: What's Victoria doing now? In an attempt to keep her mind off her baby and on her work, she consciously plugged into what she called her "mother mode," and tried to do her best to soften the rough edges of life for those around her.

An attractive twenty-nine-year-old woman with chin-length, smoky blond hair, Jewett had coveted the flight attendant's life-style for as long as she could remember. "I love to travel," she told her friends. "I love airplanes. I love airports. I love the uniforms. I love everything about flying, really."

Love was still a word she associated with flying, but a

new, earthbound passion had intervened. Jewett had been pleasantly overwhelmed by the joys of motherhood. During her maternity leave the previous year, she spent time studying child-care books, determined to be a good mother. And once Victoria was born, Jewett encountered the inevitable ambivalence of the working mother—she could not be two places at once. No matter where she was, or what she was doing, the baby was never far from her thoughts. At least this was an easy, lightly loaded flight. If she could not be at home mothering Victoria, she took pleasure in mothering her temporary charges.

Walking slowly up the aisle, she stopped to chat for a moment with an older woman, perhaps eighty, who had difficulty walking. Jewett helped her visit the lavatory, then continued on her rounds.

Three couples in her section were traveling with children, and another mother was handling two children by herself. Susan presented all the youngsters with Air Canada coloring books.

She served one more Rusty Nail to the gruff hulk of a man in the center tier of seats—the one who seemed to be a bit on edge. She knew the type. He was a tourist-class passenger who expected first-class service. She did her best, but there seemed to be no pleasing this particular passenger.

Finally she approached a stocky young man, traveling alone, who had his nose buried in the in-flight magazine.

Thus far, Mike Lord gave the flight a mixed review. The arctic trout he had for dinner was excellent, especially when accompanied by a cold beer. But the movie selection was disappointing; he had already seen Richard Pryor in *The Toy*. Nevertheless, he had just about decided to watch it again, simply to pass the time, when Jewett approached and informed him that the movie projector in this section of the airplane was not working.

20

"If you'd like, you can move up to the next section of the aircraft to watch the movie," Jewett suggested.

"No," Lord decided. "I'll just have a couple more beers and sit back and relax, eh?"

As she served the beer and moved on, Lord once more picked up the in-flight magazine. He thought ahead. He guessed that his cousin Deb was getting ready to leave home to meet his flight.

A thirty-four-year-old bachelor with his own modest home on the east end of Montreal, Lord undertook a once-a-year obligatory pilgrimage to western Canada for a summer visit with his brother Brian and his Aunt Mame and Uncle Kenny. He anticipated the journey with mixed emotions. It would be good to see his family, but he would miss his girlfriend of seven years, Danielle LaRoche, and the sanctuary of his recreation room, where a bank of stereo equipment drowned out the cares of the world and an abundantly stocked wet bar stood sentinel for the regular entertainment of his many friends. He would parlay any suggestion that the bar was for *him*, saying, in a credible impression of W.C. Fields, "I don't drink much myself, eh?"

But he had belted down several beers so far today, in part because he was never fully comfortable in the air, an attitude accentuated when, six weeks earlier, he had purchased his ticket.

Lord had driven across Montreal, through the busy heart of Canada's second largest city, on a freeway populated by suicidal tailgaters, reckless lane-changers, and ubiquitous speeders who had long disregarded the posted limit of sixty mph and now, since 1979, exhibited equal disdain for the 100 kilometers per hour limit. He had followed the signs in both French and English that delineated the contorted route off the highway and into the confines of Dorval International Airport, jammed into an industrial and residential section on the western island of

21

Montreal. Managing, somehow, to find a parking space, he emerged from his car and headed toward the terminal, and then to the Air Canada ticket counter.

Perhaps the ticket agent sensed a measure of uncertainty in the stocky customer, for she had zealously launched into a sales pitch, extolling the wonders of the particular airplane to which Lord, six weeks hence, would entrust his life. It was the brand-new, clean-as-a-whistle, super-sophisticated, state-of-the-art Boeing 767, crammed full of space-age technology. If Lord was lucky, she said, he might even get a chance to see the cockpit for himself. Many of the amenities there were the direct heritage of the U.S. Space Shuttle program. Brilliant, color-coded in-flight displays flashed the moment-to-moment status of the aircraft's operating systems as a bank of approximately sixty computers in the electronics bay beneath the flight deck monitored every nuance of flight. So modern was the aircraft that, unlike all of the others in Air Canada's fleet, the 767 computed its data in metrics, rather than the old imperial system of weights and measures. "The thing nearly flies itself!" she gushed.

"Eh, sounds good," Lord replied. But he thought: Computers? Hope the damn things aren't like ours. Lord's job as a computer operator for a telecommunications firm had been complicated lately by failures of the company's data processing equipment. He lit a cigarette and muttered, "No problem, eh?"

The ticket agent, unaware of Lord's misgivings about the age of technology, had concluded her pep talk about the 767 by proclaiming, "This is your lucky day!"

But he was not feeling particularly lucky by the time July 23 rolled around and he marched into the wide-bodied passenger cabin of the aircraft. He had fastened his seat belt securely the instant he sat down, pulling it as tightly as it would go. This thing isn't going to be a damn bit of good if we crash, he concluded, and other grisly scenarios leapt immediately to mind. After the pilot had announced a brief delay due to problems with the fuel gauges, he had begun an uncomfortable internal monologue. Is it possible for a door to fly open and be sucked into the sky? I've heard of these things hitting an air pocket and people crashing their heads into the ceiling. Better to keep the belt on all the time, eh? Looks like a great day here, but you never know when the weather will change.

He had waited impatiently for the flight to get under way, living for the moment when the no-smoking sign went off. As soon as it did, he lit up and requested a beer from the flight attendant, even though she wasn't bringing the drink cart around on this first short hop. He had barely had time to chug it down before they were on approach to Ottawa.

On the ground there, he had ignored the flight attendants' request for through passengers to remain on board, and had stepped outside in order to smoke another cigarette or two before the long flight to Edmonton.

Now they were about halfway, he reckoned. Not much more to go, eh? he thought.

Well forward of Mike Lord, ahead of a cabin bulkhead over the leading edge of the right wing, Nigel Field whizzed through a paperback whodunit. With characteristic British reserve and equanimity, he took the rigors of travel in stride. His thirty-year career with Canadian National Rail, where he had risen to the post of manager of plant capacity planning, took him on frequent trips throughout Canada.

Born near London in Hern Hill, England, Field had enlisted in the Royal Air Force volunteer reserve in 1949, earning his wings in the De Havilland Chipmunk. Part of

23

his training included a two-week tour with the Royal Canadian Air Force that expanded his sights and eventually transplanted him to North America. After graduating from the University of London with a degree in civil engineering, he planned to see a bit of the world before settling down, and chose this country as his first stop. "I'll go to Canada for a couple of years," he mused. "Then perhaps Australia and maybe New Zealand before I return to England."

In 1955 Field landed in Toronto with forty-two dollars in his pocket, quickly secured a job with CN Rail, and sent for his fiancée, Maureen, to join him. Some thirty years later they were still in Canada. The youngest of their five children was nearly on his own now, and there was a new grandchild to enjoy. Nigel Field was a happy man.

Field had long ago learned the art of the placid pilgrim. It simply did no good to become agitated by things over which one had no control. Late flights, misplaced luggage, and fouled reservations were fellow travelers on almost any business trip in this frenetic world, so if one expected and accepted such inconveniences, one would not be surprised when they occurred.

Arriving at Dorval Airport early after the hour-plus train ride from his home in Cornwall, just across the provincial line separating Quebec and Ontario, he had found himself relaxed and ready. A pleasant perk awaited him at the end of this weekend business flight. After completing his work in Edmonton, he would continue on to Vancouver for a visit with his new grandchild.

And then there was the airplane. As a former pilot, Field was fascinated by the 767.

In sum, he was a contented gentleman who had thoroughly enjoyed the flight thus far, and now happenstance offered an added surprise. Glancing up from his book, Field suddenly noticed one of the flight attendants, an

24

attractive blonde, who had been serving passengers further back but was now passing through this forward cabin. "Hello!" he called out. "Do you remember me?"

Susan Jewett stopped, studied the gentleman's face for a moment, and then brightened with recognition. "Of course!" she replied. "How are you?"

They had met several months earlier. Field had been in the midst of one of his frequent business trips. Jewett had been flying as a passenger, on her way to show off baby Victoria to her sister Judy in Smithers, British Columbia. The flight had been packed to capacity with the predictable result that Jewett, flying standby on her employee pass, was relegated to a center seat. In the confined environment, Victoria, sitting on Susan's lap, grew restless.

No stranger to fidgety babies, Field had taken the child's antics in stride. He tried to assist Jewett as best he could. She had been grateful for his efforts and a pleasant conversation ensued. At the end of the trip, Susan had revealed that she was a flight attendant for Air Canada. If Field ever spotted her on one of his trips, she said, he must make his presence known to her. She would see to it that he received first-class treatment in exchange for the gracious manner in which he had assisted her. Now that opportunity had occurred.

"Give me a minute and I'll come back and visit," she promised.

On the left side of this forward cabin, across from Nigel Field, flight attendant Danielle Riendeau prepared to eat her dinner, but was momentarily sidetracked by an elderly woman passenger who appeared tense and nervous.

"Is this perhaps your first flight?" Riendeau asked in French, guessing correctly at the woman's native tongue.

"Yes, it is," she replied, "and I'm scared to death."

"Oh, everything will be fine," Riendeau reassured. "There's nothing to worry about."

The woman hesitated, then asked timidly. "Would you hold my hand when we land?"

Riendeau explained that the law required her to sit in her assigned jumpseat near the emergency exit over the left wing. "But I'll be right here where you can see me. Don't worry about a thing." Then she smiled warmly and returned to her seat to eat her dinner.

In the air, Riendeau always made sure to get proper nourishment, choosing her foods carefully. A tall, striking French-Canadian woman with long, dark brown hair falling into casual ringlets, Danielle exuded a healthy beauty that belied the fact that, even after ten years on the job, flying still made her nauseous.

On her early flights she seemed to spend as much time in the lavatory as she did with her passengers. But she stuck out that first year, experimenting with her diet, gradually learning how to control her nausea. She learned never to fly on an empty stomach. She learned to avoid night flights and long overseas jaunts. And she learned that larger airliners were kinder to her sensitive stomach, which is one reason she appreciated the mammoth 767.

This had been an especially enjoyable month-long block of flights, made even more so by the presence of her good friend Annie Swift, who had been in a decidedly festive mood when Riendeau picked her up on the way to Dorval Airport earlier that afternoon. She had chattered about her morning of horseback riding.

Swift's mood had proved infectious. Before they knew it, the two women had dissolved into giggles at the recollection of the prank they and the other flight attendants had played the previous week upon Bob Desjardins, the in-charge flight attendant. They had learned that it was

Desjardins' birthday. In the air, somewhere over Saskatchewan, after applying copious amounts of bright, garish lipstick, they had summoned the reserved young man to the galley of the airplane and smothered his face with kisses. Then, as they sang a cheerful chorus of "Happy Birthday," they had paraded him through the aisles of the airplane to the delight and amusement of all the passengers. Finally they had fashioned a makeshift birthday card, reapplied their lipstick, and kissed the card, taunting Desjardins to guess whose lips were whose. He had failed the test, to the delight of the other flight attendants.

After she had cleared away dessert and prepared the passengers for the movie, Susan Jewett hurried forward for her promised visit with Nigel Field.

The British-born businessman put down his book as she approached.

"Are you going to watch the movie?" Jewett asked.

"No, I'd rather chat with you," Field replied.

Jewett knelt on the seat in front of Field, facing backward. Field inquired about the baby—Susan's favorite topic—and she thanked him once more for his patience and kindness during their previous flight together. "Victoria's a year old now!" Jewett beamed.

Having opened up this line of conversation, Field now found himself deluged with details of Victoria's babyhood. There were to be no processed foods for Victoria, only hand-prepared, unsprayed fruits and vegetables from the garden. Victoria's stuffed animals contained no toxic perfumes. Dr. Spock was currently Susan's favorite author. She confided to Field that she and her husband (an Air Canada 727 first officer) had recently come to a decision that would alter their lives considerably. They would give up Susan's salary, at least for a year, and Susan would forsake the glamour of the sky for full-time motherhood. Her mind was on the future now—all she had to do was make it through the next few months, she said, and then she would begin her year's leave of absence.

Field countered with his own family news. He boasted of his new grandchild, whom he would see on this trip.

Jewett and Field visited quietly for about ten minutes before the flight attendant began to wonder how she could extricate herself from the conversation, lest her colleagues think she was ignoring her duties. She suddenly brightened, remembering that Field was a veteran pilot of the RAF reserve. She asked, "Would you like to see the flight deck of this new plane?"

Field was delighted. "Indeed I would," he replied. "I'd love that."

Such an excursion was unthinkable on a U.S. airliner, where FAA regulations allow no visitors in the cockpit. In Canada, with fewer security problems, the rules are more flexible, and this was the perfect way to repay a previous kindness. What's more, Susan knew, there was no finer cockpit host than Captain Pearson. Not only was he a great pilot, but he was Mr. Personality as well. "With Pearson, you know you're going to get a show," Jewett would tell her passengers. Children and adults alike were mesmerized by the force of Pearson's charisma and joie de vivre. Now she could take advantage of Pearson's magnanimous nature, to his delight as well as theirs. This is great, she thought. Nigel will love it.

"You have to come up front," Jewett beamed. "This pilot is such a great guy! A character, eh? Ask him anything."

At that instant, both Jewett and Field noticed a sudden, gentle drop in altitude. It caused neither of them concern; there were a thousand reasonable explanations for an airliner to change altitude: course corrections, weather, other traffic, changes in the pattern of the prevailing winds. But they glanced out the window, unconsciously monitoring the pilot's action. They could discern no message in the light cloud cover below, and they quickly dismissed the incident from their minds.

Susan excused herself to go to the cockpit to clear the way for Field's visit. She moved toward the front of the airplane and disappeared past a bulkhead.

Seated on the left side of the aircraft, over the wing, passenger Shauna Ohe felt the same sudden dropping sensation as Jewett and Field, but she reacted with alarm. She sat up straight in her seat. "What's going on?" she asked her companion, Michel Dorais.

"Nothing," he said, glancing up from his history book.

Ohe and Dorais were proof of the axiom that opposites attract. At age forty-four, Ohe was nine years older than Dorais. A divorced mother of three grown children, Ohe's impulsive and capricious nature was in direct contrast to Dorais' more pragmatic approach to life. She came from central Saskatchewan, from a small village called Wynyard, established by nineteenth-century Icelandic settlers. She had married and settled in Edmonton, but after the break-up of her fourteen-year marriage, seeking change, she had searched for *the* job that would bring new direction to her life. She found Dorais instead.

Balding and bearded, Michel Dorais had achieved success in life through the rational application of hard work and tenacity. He had emigrated from his native Montreal to Edmonton in 1969 and immediately set about to overcome the effects of prejudice arising from his French-Canadian heritage. He took a job as a shopping center handyman, working nights so that he could study English during the day. He segregated himself from the small

French-speaking population of western Canada so that he would be forced to learn English. He read a book a day —in English.

Gradually the effort paid off. His language skills improved enough so that he could launch a career as a life insurance salesman and, by 1983, was a highly respected professional in his field, a member of the insurance industry's most elite group, the Million Dollar Round Table, and served a clientele that included some of Edmonton's most successful executives.

In 1979 Shauna Ohe walked into his office and applied for a job. It was not love at first sight. Ohe did not even accept the job, but the two struck up a friendship that remained platonic for several years. Only recently had they become lovers and, even more recently, committed their lives to one another.

On this day they were returning home from Montreal, from a visit with Dorais' parents and a side trip driving through the Maritime Provinces, and they had boarded the aircraft pleasantly relaxed and ready to return home.

Their contentment had been disrupted while still on the ground in Montreal. The pilot explained over the intercom that takeoff would be delayed due to a problem with the fuel gauges.

"Why can't they just fill up the plane and go?" Ohe had muttered.

"They can't do that," Dorais decreed.

As it happened, Dorais had a friend who was a load control agent for Air Canada, and he had once detailed for Dorais what is known as the "minimum fuel concept." Now he explained the process to Ohe.

Fueling a jetliner is a bit of an art as well as a science. Quite obviously, no pilot is willing to take off with an insufficient load of fuel, but the converse is also true. Kerosene stored in the wings is a heavy burden. Too much fuel is a safety hazard, requiring a lengthened takeoff run. It is also inefficient and therefore costly. The idea is to calculate the fuel load so that neither safety nor efficiency is compromised.

The minimum fuel concept precluded any easy solution to the problem. "It would be expensive to fill the tanks full," Dorais had explained to Ohe. "If the pilot requested that he would be in shit."

Now, hours later, when Ohe expressed alarm over the sudden drop in altitude, Dorais was again ready with a logical explanation. He reminded her of an incident they had experienced during a recent flight from Victoria to Vancouver. For some never-explained reason, the pilot had cut back power to the engines, and the sudden deceleration was noticeable. Ohe had panicked, and Dorais did not want her to do so again.

"Remember the other time," he counseled. "There was nothing to worry about then and there is nothing to worry about now. Just relax."

Mike Lord was sipping his beer when he felt the seatbelt tighten across his midsection, the same sensation as when a speeding roller coaster clears the crest of a sharp hill. He glanced around anxiously, and his eyes met those of a young woman sitting two rows behind him in the center tier of seats. She shrugged her shoulders. He shrugged back. Who could tell what these pilots—and their computers— were up to?

A pad of stationery on her lap, flight attendant Annie Swift sat in the middle tier of the final row of seats. Having finished her dinner, she was penning a short thank-you note to her boyfriend's sister, who had been their hostess during their recent sabbatical at Premier Lake, a resort in Okanagan, British Columbia. She told her how the visit had brought her back to her first love of horseback riding.

"I rode Nadjaya this morning," she wrote. "Ten years, can you believe it's been ten years since I've ridden her?" Nadjaya, the Russian name Swift had bestowed upon the bay thoroughbred, translated as "hope and expectation."

To sit astride a horse, to control the power of this exquisite animal, had once been Swift's all-consuming passion. Showing horses, caring for them, riding cross-country, and instructing others to ride had been her lifeblood. Why had she given it up? she wondered. How could she have? But she knew that the answer was intertwined with her personality. Whatever Annie Swift determined to do, she *must* do 100 percent. When she joined Air Canada as a flight attendant a decade ago, she had focused her energies on her job. She could not bear to become merely a recreational rider. "It would diminish me," she had said, "it would just be too painful." And the exhilaration of riding had faded into memory, until this very morning.

This year had been a difficult one for the thirty-one-year-old flight attendant. She had decisions to make about her life and her future. Her relationship with her boyfriend, an Air Canada second officer flying Boeing 727s, had reached that crucial turning point where both had to decide to move forward or to back off. He was a wonderful man who cared for her deeply, but Swift had found it difficult to respond to his desire for a permanent commitment from her. The vision of domesticity and a white picket fence was at once appealing and terrifying to her. Although she knew some elusive "something" was missing from her life, she was unsure that marriage was the answer.

The trip to Okanagan was an attempt to sort out their future. Swift hoped that the spectacular vistas of the Canadian Rockies might provide perspective on her life. She had found her answer, not in the mountains, but on the prairie. Each day she had thrilled to the sight of herds of wild horses, running free, and she knew. *I must get back to this.*

The day she had returned to the house she shared with her boyfriend in Pointe Claire, Quebec, only five miles from Montreal's Dorval Airport, Swift had called her close friend and surrogate uncle, Ludwig Popiel, to whom she had sold her beloved Nadjaya, and had made arrangements to ride again.

This very morning, when she had climbed upon Nadjaya's strong back and set off upon the meandering trails of the French-Canadian countryside that crisscrossed the land, unobstructed by walls and fences, she knew what had been missing in her life. "This is where I belong," she said with conviction.

Riding was the one passion in her life that burned brighter than flying, and she savored the memory of the morning even as she soared over central Canada.

Her reverie was interrupted by a sudden tug, an unmistakable sensation, as though the aircraft had nosed upward slightly, nearing a stall. A sense of déjà vu encompassed her. Only a few months earlier, on a Lockheed L-1011 flight, a computer error had caused the plane to stall and lose altitude quickly. A decompression followed, alarming everyone. The pilots had caught the computer error immediately, but they were forced to descend rapidly in order to reach a safe breathing altitude.

The emergency had left a lasting impression. Swift knew that the sensation was one she would always remember with trepidation—and now she had experienced it again.

3

The Cockpit

In front of each pilot, an amber warning light glowed, redundant signals adding their alarm to the buzzer. Reacting instantaneously, First Officer Maurice Quintal checked a lighted message on the video screen of the Engine Indictor and Crew Alerting System (EICAS) in front of him. This detailed the source of the problem. "Something's wrong with the fuel pump," he reported.

"Left forward fuel pump," Captain Pearson confirmed.

Quintal pushed the amber warning light to deactivate it.

Without hesitation the two men reverted to a standard operating procedure, pounded into them by years of training and retraining and honed by experience. At the sound of the buzzer, routine yielded to the extraordinary. Whether the situation proved to be critical or merely bothersome—or even if it was a false alarm—whether its duration would be extended or brief, until it was over the two pilots would confirm every iota of data, double-check every action. This was a moment when a pilot really earned his pay, when a buzzer and an amber light suddenly signaled that one of the world's most sophisticated engineering marvels was in need of help from a human being.

First, Pearson thought, define the problem.

"Okay, what have we got here?" he asked. "I hope it's just the fuckin' pump failing, I'll tell you that." On each engine a pump pulled fuel from the tanks. In each of the three fuel tanks—one in either wing and one in the belly—two fuel pumps also pushed a constant supply of kerosene to the engines. Whether the aircraft climbed or descended or banked left or right, at least one of the pumps in each tank would be in position to feed fuel to the engine pumps. The entire system was harmonized by an interlaced network of piping. If necessary, the pilots could cross-feed fuel among the tanks.

The amber warning light told them that the forward pump in the left wing tank was laboring at abnormally low pressure. Why was not their immediate concern. The pilots' first priority was to circumvent the trouble spot. Keep the aircraft flying. Pearson opened a cross-feed valve so that the left engine could be properly fed from the right wing tank until he and Quintal could determine their next action.

The Quick Reference Handbook was still open on Pearson's lap. He flipped through it rapidly, found the appropriate section, and read aloud to the others: "If one main tank pressure light is on, continue normal operation. If one center tank pressure light is on, cross-feed valve switch on." He concluded, "So we don't need it." According to the handbook, since only one of the left tank fuel pumps was operating at low pressure, there was no cause for alarm. "That's it," Pearson declared. "Continue normal operation." He closed the cross-feed valve and for a few moments the cockpit grew quiet.

36

But regardless of the mollifying information in the Quick Reference Handbook, each of the three men in the cockpit quickly, silently reviewed the events of the day. The straightforward logic of the manual could not take into account what these three men had seen and heard in Montreal and Ottawa as the aircraft was serviced. In retrospect, the events of the day were disquieting. A vague sense of unease permeated the cockpit as each man considered the notion that what had just occurred might be merely a prelude.

Dion watched the amber warning light flicker off and on and then glow steadily, and he searched his memory. From time to time a mechanic must drain a fuel tank prior to working on it, and Dion knew that this particular alarm could be expected shortly before the tank went dry. On the other hand, it could merely indicate that the pump was clogged. Dion recalled a case in which a mechanic had inadvertently left a rag in the fuel tank, and it had worked its way to the pump inlet, jamming the system.

Thus, the message of the buzzer and the amber warning light was ambiguous, upsetting these three men over a minor problem, or alerting them to the possibility of catastrophe.

Only a few flicks of Pearson's digital watch passed before vague anxieties became fears of substance. Four more warning beeps blared through the intensified atmosphere of the cockpit. A new message appeared on the EICAS screen, bearing the ominous information that the *second* pump in the left wing tank was now failing.

"Oh, fuck!" Pearson muttered, making an instantaneous decision. "We've got to go to Winnipeg." This second failure was simply too remarkable to be a coincidence. Pearson had to believe that the pumps were failing due to related reasons. It was obvious now that the left side fuel system was experiencing a major problem, and the captain's decision was immediate. We've got some computer problem that I don't understand, Pearson thought. I'm going to Winnipeg to get it fixed. He would divert to the nearest suitable alternate landing, Winnipeg, southwest of their present position.

Remembering that he had an expert mechanic in the cockpit, Pearson asked Dion for his assessment. What was the meaning of the failure of both left wing fuel pumps?

"Ah, you could, my own personal thought, you might be low on the left fuel tank," Dion said. He, too, believed that the failures were related. The problem was not the pumps; rather, something was wrong with the fuel supply. Dion reasoned that someone must have erred in calculating the fuel load of the left wing tank. This being the case, they must now activate the cross-feed system and fuel both engines from the right wing tank. Given the uncertainty of the situation, Dion concurred with the decision to divert to Winnipeg.

What happened? Quintal thought. His mind raced back to the uncoordinated refueling procedure in Montreal. He recalled the difficulty the mechanics had in performing a simple arithmetic function in order to calculate the fuel load.

Pearson interrupted the first officer's thoughts, repeating his decision: "Let's head for Winnipeg *now!* Get a clearance direct Winnipeg." Even in the heat of the moment Pearson had the composure to add, "Please."

Although Quintal had been piloting the airplane on this leg to Edmonton, rank and experience clicked into place automatically, wordlessly. Pearson disengaged the automatic pilot and took the controls. He would hand-fly the remainder of the aborted flight. Quintal knew that his job was to do whatever he could to help. Behind them, Dion watched attentively, ready to offer what assistance he could. But he had more on his mind. Of the three, he was the only one with his family on board.

Quintal grabbed his microphone in order to request immediate clearance to divert to Winnipeg. But he hesitated for a moment before he spoke into the microphone. I know what is going to happen, he thought. He tried to put out of his mind the nagging realization that he was aboard this flight only by a quirk of fate. Paul Jennings was the scheduled first officer on this month's block of flights with Pearson, but he was ill today, and Quintal was first on the standby list. His seniority status was too low to provide him with a regular monthly schedule. On reserve, like a physician on call, he had to be prepared on a moment's notice to head off to San Francisco or Barbados or London or Paris or Geneva or Dusseldorf or Edmonton. So he was here now, instead of Jennings, and suddenly confronted with an emergency of unknown dimensions.

It was the first moment of Quintal's life when he did not want to be in the air.

With the possible exceptions of Wilbur and Orville Wright, no human had labored more tenaciously to fly than Maurice Armand Nicolas Joseph Quintal.

After high school he enrolled at Montreal's Institute Aeronautic de Quebec, choosing the school because of the mystique and the lure that the word "aeronautic" held for him. He planned to become an airplane technician, knowing that he could earn a better living in that line than by working in his father's butcher shop.

At the Institute, however, he found not a living but a life. The slight, brown-haired, soft-spoken French Canadian made friends easily, and it was with a group of his fellow students that he had taken his first ride in an airplane, a small single-engine Cessna 172. He sat in the back and found his heart thumping with excitement as the pilot revved the engine and taxied the plane down the runway. The very moment that the airplane lifted off the ground, magically overcoming the force of gravity, Quintal's quest began. He would not work on airplanes, he would fly them.

This resolve carried him for a year and a half as he continued his technical training at the Institute during the week and worked weekends at the butcher shop to pay for flying lessons. He gradually realized, however, that his fantasy was too expensive for his reality. He wanted his private license, his commercial license, his instrument rating and his twin-engine rating. He wanted to fly for the airlines, but at the rate he was going, it would take forever. There seemed to be only one viable alternative, and that was to join the Royal Canadian Air Force and let the government teach him to fly.

From the beginning, Quintal realized his mistake. He joined the RCAF in April 1968 almost concurrently with Prime Minister Pierre Trudeau's decision to de-emphasize the military, cut the budget, and close training schools. During Quintal's stint, the Canadian military census dropped from 120,000 to 85,000. For his first assignment Quintal was sent to St. Jean, Quebec, not to learn to fly, but to study English.

Nevertheless, he eventually found himself flying for a scaled-down RCAF, first in the Chipmunk trainer, then the CL-41 jet trainer (known as the Tutor), the jet fighter used by the RCAF aerobatic team, The Snowbirds.

Wherever the young officer was stationed, he had little patience with the military mentality. Try as he might, Quintal could not see how the sophomoric behavior of his "superiors" related to the production of fighter pilots for Her Majesty's defense. He spent far more time exorcising dust from his barracks than soaring through the skies. He had joined the Air Force to fly, not to clean his room, so he parted company with the RCAF at the earliest opportunity, one more casualty of the prime minister's campaign. "I left the Air Force to *become* a pilot," he would say.

In November of 1968 Quintal was a civilian once more. He could already fly. *He* knew that, but bureaucracy required him to prove it by logging sufficient hours, and he was woefully short. He sold his car, borrowed money from his father, and set off to fill his logbook with the necessary hours.

Rental planes in Montreal cost twenty dollars per hour, Canadian, and that was too much. In the States Quintal could fly for the equivalent of eleven dollars per hour, so one Saturday he drove across the border to Plattsburgh, New York, rented a plane for the duration of a clear, beautiful weekend and headed south, ending up in Greensboro, North Carolina. At other times he ran newspaper ads undercutting commercial fares to Nassau by fifteen dollars, and, if he booked enough passengers to cover expenses, flew them off on holiday, certifying the hours in his logbook. He learned of a pilot who was planning a trip to West Palm Beach, Florida, talked his way on board, and flew the borrowed Cessna 120 to Miami, around the Gulf of Mexico, down to Yucatán, and back. Anything to log more hours.

Quintal managed this gypsylike existence for two years, holding down a series of odd jobs while grabbing the stick of an airplane whenever possible. Eventually he accumulated enough hours to qualify for an instructor's rating. Working part-time at a flying school, he finally earned his first real money (outside the RCAF) as a professional flyer, but he was still a long way from the goal that spurred him on whenever he gazed overhead at a jetliner taking off from nearby Dorval Airport.

A friend mentioned that he knew of a man who was

hiring pilots to fly DC-3s up north. "You have an instrument rating, haven't you?" the friend asked.

"Yes," Quintal said, not mentioning that his instrument rating had expired.

Quintal was hired on the phone and had to cram for—and pass—the required written exam. He scraped together enough cash to rent a twin-engine Apache and persuaded a Ministry of Transportation inspector, who happened to be a friend, to give him a check ride on short notice. After flying around Montreal for about an hour, Quintal bounced the Apache in for a landing and his friend said, "Well, technically you failed, but I'll sign your license anyway. You're a little rusty, but you only need a few more hours."

That very day Quintal headed north. He spent the next winter flying cargo on and off the frozen Matagami River and he flew a regular passenger schedule during the summer. The hours were long—from seven in the morning to seven at night, 120 hours a month—but the pay was good; more importantly, when he returned to Montreal, he had acquired enough expertise to land a job as a flight instructor for a provincial pilot training program for college-age students. Still, the commercial airliners passed overhead, controlled by the privileged few.

In 1973, Gilles Lafreniere, a recruiter from Air Canada, came to interview Quintal's students, seeking prospective pilots. Quintal learned of the visit, but he suspected that his boss wanted to keep him away from Lafreniere. That evening, as Lafreniere and the instructors congregated in a local bar, Quintal awaited his chance. He followed Lafreniere to the washroom and, standing next to him at the urinal, asked, "Do I understand you are here to recruit our students?"

"Yes."

"Hey, how about me?"

Two weeks later he was in ground school, learning his new job as a second officer on the DC-8. There followed four years on the DC-9, three on the Boeing 727, and a year on the L-1011. In 1982, like Captain Pearson, he was one of the first Air Canada pilots to volunteer for training on the Boeing 767, and by July 23, 1983, he was a three-month veteran of the new aircraft, with seventy-five hours of flight time on it.

On the outside, Maurice Quintal was a man who had achieved success, living in a fashionable, contemporary home in the northern Montreal suburb of Lorraine with his wife and two young sons. But there was rough weather in his life that had not been in the forecast. His wife suffered from a chronic, debilitating illness, and the pressures kept Quintal in a constant state of tension. He loved to fly; he needed to fly. On call, with an uncertain flight status, he accepted every available trip.

Yet he was also needed at home, by his wife as well as eight-year-old Jean François and five-year-old Martin.

It was a demanding, dual personality role, and it gave rise to the expected ambivalence when, on July 23, 1983, the call came from Crew Scheduling. Paul Jennings was ill. Would Quintal take the first officer's seat on Flight 143?

He accepted, for the usual reason of garnering the flight time, but also because he craved the solace of the sky on this beautiful summer day.

One final fact had capped his pleasure at the summons. He would be flying with Bob Pearson, a bit of a legendary figure among Air Canada pilots. Pearson and Quintal had flown together only a few times, but the younger man had reason to hold Pearson in high regard.

A few years earlier, a fire had gutted the ground floor of the apartment building where the Quintal family then lived, causing considerable smoke damage to their upstairs flat. Quintal had moved his family out to temporary quarters and was in the process of cleaning and painting his soot-filled home when Crew Scheduling called and asked him to substitute for another pilot on a four-day flight cycle to Winnipeg and Windsor. Quintal hesitated, but the scheduler was insistent. The airplane in question had already been at the gate for an hour, waiting for a first officer—any first officer. Quintal reluctantly agreed to take the flight.

On the final leg of the trip, an inspector from the Canadian Ministry of Transport boarded the aircraft in Winnipeg to conduct a routine license check. In the mayhem of the fire and the last minute assignment, Quintal had neglected to have his medical license revalidated; it had expired at midnight the night before. The inspector was sympathetic; nevertheless, regulations prohibited Quintal from completing the flight. Another first officer was called in to take over.

Upon returning to Montreal, Quintal phoned his boss and explained what had happened.

"You're going to get a reprimand," was the terse response.

Quintal was incensed. He had done the company a favor by agreeing to take the flight at the last minute, in spite of his family emergency. That is when Quintal met Pearson, the representative in his area for the Canadian Airline Pilots' Association. By telephone Quintal tracked Pearson down at the local ice hockey arena, waited while he was paged, and told his story. Within the hour Pearson met Quintal at the airport, ready for battle.

After a week of tough negotiations, Pearson was able to get Quintal off with a slap on the wrist in the form of a two-week suspension that enabled him to keep his job as well as his seniority.

Quite apart from appreciating the captain's superb flying ability, Quintal just liked to be in the same cockpit with the man who had saved his career.

On the afternoon of July 23, 1983, the first officer hurriedly packed his overnight bag, bade good-bye to his troubled family, and drove to the airport, anticipating the incredible upward pull of a 767 racing down a runway, picking up sufficient speed so that the configuration of the wings forced the air to pass over them far faster than under, creating a low pressure zone above the wings that sucked the vehicle into the sky. The tingling phenomenon of lift was more pronounced on the 767 than on any other plane Quintal had ever flown. The 767's rate of climb was so much steeper than an ordinary jetliner's that it sometimes produced gasps of alarm from passengers—and occasionally from "dead-heading" pilots, journeying to their own flight assignments.

Other pilots were envious, a realization reinforced when Quintal had arrived at the Air Canada flight planning office and encountered his good friend and neighbor Gilles Sergerie, a first officer on the Boeing 727. Sergerie was scheduled to take off at 6:30 P.M., an hour later than Quintal.

"I didn't know you were scheduled today," Sergerie said.

"I wasn't. I was on call. Paul Jennings is ill. I'll be taking 143."

"More time on the 767, yes?"

Quintal had flashed an impish grin.

Maurice Quintal was not grinning now, as he gripped his cockpit microphone.

"Winnipeg Center, Air Canada 143," he called.

"Air Canada 143, go ahead," came the reply.

"Yes, sir," Quintal said. Then he spoke the four words

that would carry, not only to Winnipeg ATC, but to aircraft cruising over a wide radius. Pilots throughout the skies of central Canada now pricked up their ears. Gone was the droning routine of the early Saturday evening. Crew members in other cockpits turned to look at one another, as they heard Quintal's voice crackle over the radio: "We have a problem."

4

Winnipeg ATC

Fifty-year-old Ronald James Hewett, a twenty-two-year veteran of the Canadian air traffic control system, was on duty at the Winnipeg Air Traffic Control Center, located on the second floor of the administration building of Winnipeg International Airport. He was monitoring a long-range radar screen that covered a circle 200 miles in diameter, centered upon Red Lake, Ontario, when the call came in. "We have a problem," said a distant voice with a French-Canadian accent. "We're going to, uh, requesting direct Winnipeg."

Hewett was working with two of three radar systems available to him. Primary radar, which simply bounces a signal off of any object in its range and returns it as a blip on the screen, was, by 1983, nearly phased out of the world-wide air traffic control system. In fact, of the eight radar stations that Hewett could monitor on his screen, primary radar was only available from Winnipeg.

At this moment he was monitoring the signal from a radar station near Thunder Bay, Ontario, that used two more sophisticated systems, both dependent upon operational equipment on board the aircraft he tracked. On Flight 143, and on all other jetliners, was a transponder that responded to the radar signal by sending back an electronic transmission encoded to identify its source. Hewett was also monitoring the new digital system, which was able to receive more complete flight information via Flight 143's transponder signal, feed it into a computer, calculate the altitude and ground speed, and display the data within a triangle superimposed on the radar screen over the target.

Thus, Hewett was able to identify Flight 143 immediately among the handful of blips on his screen, check its altitude and speed against other traffic in the area, and issue clearance for the stricken aircraft to begin its descent to Winnipeg.

Only four seconds after hearing Quintal's call Hewett replied, "Air Canada 143 cleared present position direct Winnipeg. We're landing on runway thirty-one. You're cleared to maintain 6,000 descent your discretion."

Hewett alerted floor operations supervisor Warren Smith and the word quickly spread that an emergency was in progress. Communications from Flight 143 were switched from Hewett's headset to a direct speaker system, so that all the controllers in the room could hear. Len Daczko, who was assigned to control traffic on approach to Winnipeg, was the most directly concerned. He must be prepared to take over when the aircraft closed to within about thirty-five miles, guiding it into its alternate airport despite any unforeseen complications arising from this undefined problem. Hewett and Daczko heard Quintal acknowledge: "Thank you. Cleared to Winnipeg 6,000."

Working quickly, Hewett passed the responsibility for other aircraft in his sector to other controllers in the room—two or three to west radar, one to Thunder Bay sector—and he gave another over to arrival controller Daczko earlier than he usually would. Now he could focus all his attention on Flight 143. He was thankful traffic was light. Administrative supervisor Steve Denike joined Smith, standing behind Hewett, watching, listening, waiting to learn the extent of the danger faced by the voice that reverberated through the alerted room.

5

The Cockpit

Having received from Hewett an immediate pathway to descend to 6,000 feet on approach to their alternate airport, Pearson and Quintal set about their business quickly, shunting their misgivings to the sidelines as best they could. Pearson put the aircraft into a gentle left turn, heading it south-southwest, toward Winnipeg.

The pilots' training enabled them to concentrate upon the immediate reality. There was only one objective now: do everything in their power to bring this aircraft and its passengers safely back to earth. Whatever the cause of the problem in the left fuel tank, they would worry about it later. It was the effect that concerned them now. They hoped to land with power remaining in both engines, but if the port side engine failed, as now seemed possible, the task would be complicated. Nevertheless, it was the kind of situation they were trained to handle.

Flight 143 was now 128 miles north of Winnipeg. At 1:14 GMT, Pearson began the descent that would take them from 41,000 feet to 6,000 feet. He throttled back the

51

engines, producing an immediate, noticeable response, as though someone had applied the brakes. Losing forward speed rapidly, the aircraft began a precipitous descent, the initial portion of a landing pattern that is simply a controlled fall from the sky. Although a pilot has many controls at his disposal to effect a landing, including air brakes that rise from the upper wing surfaces, flaps that descend from the trailing edges of the wings, and slats that protrude down from the leading edges, his primary control is engine speed. On a normal approach the engines are set at or near idle, providing only enough thrust to counteract the drag caused by the enormous engine pods themselves.

In this very special case, reducing engine speed not only put them on approach to Winnipeg, but provided the additional benefit of conserving precious fuel.

Working quickly, efficiently, Pearson and Quintal programmed the Horizontal Situation Indicator, one of the many computer-driven cockpit monitors, to produce a pictorial display of their descent profile to Winnipeg. In bright graphics, computers portrayed the proper glide path that would bring Flight 143 to the threshold of runway 31 in Winnipeg.

Seeing how busy they were, mechanic Rick Dion asked if he should leave.

"Uh, no, why don't you stay up here and see if you can be of any assistance?" Pearson requested.

"All right. Okay."

The captain calculated the descent profile so that they would arrive over the Winnipeg airport still more than a mile high, in order to give himself plenty of maneuvering room. As he homed in on the heading to Winnipeg, he vowed, "I'm not going to take any chances."

"No, 'specially with the quantity out," Dion agreed, referring to the previously unspoken issue that was on all their minds. The fuel quantity processor was useless. Thus, there was no way of assessing the fuel supply with certainty. The flight management computer showed about 11,000 kilograms of fuel remaining on board, but that figure was based upon human input, not automatic measurements. According to the computer, about half of that load was in the left wing tank. If that information was accurate, why were the pumps failing? The discrepancy between the computer data and observable reality caused the three men to make a tacit decision that Pearson would enunciate as: "to hell with the computer." For the duration of this flight the pilots would rely more upon their own human skills and instincts than upon dubious electronic data.

Searching for every spare ounce of fuel for the starving left engine, Pearson switched on the center fuel tank pumps to make use of any kerosene that may have remained there from previous flights or, as sometimes happens, had dripped in from the wing tanks.

The cockpit was quiet, tense, as Flight 143 sank deeper into the late afternoon sky toward the cloud cover below. A mere five minutes had passed since the initial warning beeper had alerted the pilots to the problem with the left fuel tank pumps.

Suddenly four more beeps sounded a further ominous warning. This was followed within seconds by another four beeps. A total of six amber lights now glowed on the control panel, bearing an eerie message that sent icy spasms along the spines of the three men in the cockpit. The problem was no longer confined to the left fuel tank. Now, all six pumps serving all three tanks were failing. Whatever the cause, the effects were intensifying rapidly.

"Goddamn, they're all going out," Pearson barked. *"Get Bob up front!"*

6

The Purser

Bob Desjardins, the in-charge flight attendant, or purser, had experienced a bit of difficulty in settling the passengers down for the movie. For some unknown reason the projector had malfunctioned in the rear cabin. Unable to fix it, Desjardins had announced that passengers who wished to watch the movie could move forward. Several had done so. Others had grumbled epithets at Air Canada and remained in place. It was the kind of operational snafu that occurs on any airliner and, it seemed, more often on a new one. Desjardins had noted the problem in the cabin log book so that mechanics in Edmonton would check it out.

Finally, he found time for his break. He heated his dinner and sat down at his post on the left side of the forward cabin, just aft of the flight deck, to enjoy his second steak of the day. It would not be as good, he realized, as the one his wife Elaine had prepared earlier that afternoon on the charcoal grill that stood on the deck of his backyard swimming pool.

Trim, neat, classically good-looking, and quietly self-assured, a dark mustache accenting his deeply tanned face,

Desjardins was naturally shy and soft-spoken, but had cultivated a presence that inspired confidence in even a faint-hearted flyer. If he looked more like a pilot than a flight attendant, it was not coincidence, for he was both. He had been an Air Canada flight attendant for a dozen years, but he was also a licensed pilot in his own right, with over 3,000 hours in his logbook. He held an airline transport rating license in Canada, a senior commercial license in the United States, and an instrument Class I rating. He often piloted charters, usually twin-engine turbo-props, as well as a wide variety of single-engine planes. Had he started his flight training a bit earlier, he would likely have sought a career in the cockpit, but by the time he had investigated that option he was already thirty-one years old, a bit long in the tooth—by airline standards—to sign on as a second officer.

Nonetheless, his experience and knowledge made him a welcome addition to any flight manifest. Pilots felt good about having him in charge of their passengers; other flight attendants liked and respected him.

That afternoon, basking in the warm sunlight next to his pool, enjoying the aroma of steaks on the grill, together with Elaine and their toddling two-year-old daughter Julie, Desjardins had experienced the euphoric realization that life doesn't get much better than this. There was even a new baby on the way. I am one lucky fellow, he had mused. He said aloud to Elaine, "They should cancel this flight. This is much too great a day to spend working."

"Never happen, Bob," she countered. "Too many people like to travel on days like this."

"I suppose," Desjardins had concurred.

Elaine's statement was belied by the passenger manifest of Flight 143, for few people had chosen to fly to Edmonton on this sparkling afternoon. Thus the workload was light, and Desjardins was relaxed and unhurried as he began to eat.

Before the purser could swallow the first bite, however, he saw the face of mechanic Rick Dion peering around the corner of the narrow aisle that led from the first-class passenger cabin to the flight deck. "The captain would like to see you," Dion said. "He has a problem."

Desjardins was too much the seasoned flyer to become alarmed over the word "problem." It could mean almost anything, from a malfunction in the intercom system to a complaint about the dinner. Whatever it was, he would address it. Accustomed to interrupted meals, Desjardins pushed his dinner tray aside and followed Dion forward and onto the flight deck. "Yes, Captain," he said.

It was Quintal who answered. "We're going to Winnipeg," the first officer announced. "We think we have problems with our fuel system. We're diverting to Winnipeg. We're presently 120 miles . . . it'll take about twenty minutes."

"Get your flight attendants and brief them for an emergency landing," Pearson interjected. The captain stressed that he did not want the passengers alarmed; he only wanted the flight attendants prepared in case it became necessary to institute emergency landing procedures.

Desjardins left immediately, his mind spinning. This was more of a "problem" than he had anticipated. A part of him longed to remain in the cockpit and take charge of the aircraft himself, but that was the fantasy of a frustrated commercial pilot. He did not have enough information to comprehend the scope of the crisis, nor time to linger and find out more. *Problems with our fuel system! What did it mean?*

It was not so much what Pearson and Quintal reported that bothered Desjardins. It was, rather, the way they said it. Pilots cultivate a matter-of-fact tone whenever they are forced to report "problems." They do not wish to raise undue alarm, nor cause panic. Some, quite consciously, always affix the word "minor" to the front of the word "problem." But at 41,000 feet, could any problem with the fuel system be minor?

Desjardins had seen the cockpit ablaze with amber warning lights—enough to indicate that more than one operational component was affected. The narrow aisle leading from the flight deck back to the passenger cabin now seemed to close in about him. He ordered himself to remain calm, to take all possible precautions to protect the passengers, to do his job and let others do theirs. Of this Desjardins was sure: If he had to be encased in the now-claustrophobic cabin of a wounded airliner, the man he would most want in control, other than himself, was Bob Pearson.

As Desjardins reached the end of the narrow passageway leading away from the cockpit, flight attendant Susan Jewett attempted to enter, but the in-charge blocked her way. Assuming that he was being playful, Jewett smiled and asked, "Are you going to move?"

Desjardins was humorless. "No, no, go to the back. . . ." he said.

"I just want to bring a passenger in," Jewett explained.

"No," the in-charge countered. "Go to the back. Get your book. We have problems and we're going to land in Winnipeg."

As Susan stared at Desjardins, irritation preceded fear. First the delay in Montreal, now this, she thought. What next? She turned on her heels and strode quickly toward the rear of the aircraft, back toward Nigel Field, sitting on the right side of the aircraft, not quite halfway back. The cabin was darkened, illuminated in relief by the flickering lights of the movie. As she approached, Field tugged at his seat belt, posturing himself to rise and join her for the cockpit tour she had promised to arrange.

Jewett held up her hand and motioned him back down. Her features carried a strange expression, one that Field

58

found difficult to read. Reality was taking hold of her, provoking anxiety. Flying, to her, had been a decade-long adventure, a worldwide jaunt through smooth skies on trustworthy, comforting machines. She did not know—could not know—the seriousness of the situation, but she took her cue from Desjardins' uncharacteristically curt order. Now the color had drained from her face. She wrung her hands anxiously.

For an instant, Field thought, she seemed to stare through him. Then she leaned forward and whispered conspiratorially, "Don't tell anyone, but we have a mechanical and we're going to have to land in Winnipeg. . . ."

Field's calm, British feathers were barely ruffled. "What's the problem?" he asked. He muttered the only possibility that came to mind. "Undercarriage?" Even as he said the word, he realized that it made no sense. If an airliner experienced problems with its landing gear, that issue would arise closer to the destination, not while cruising at 41,000 feet over Red Lake, Ontario.

"I don't know," Jewett replied, still rubbing her hands together. She disappeared down the aisle and into the galley.

Forward, in the first-class passenger cabin, Desjardins rifled quickly through his briefcase, located his emergency manual, opened it up to the 767 section, and removed the card detailing emergency landing procedures. Then he attempted to assume the same bearing of bravado practiced by the pilots. He strode briskly down the aisle of the tiny, nearly empty first-class cabin, trying to gauge his speed for maximum efficiency without putting anyone on alert. He passed a bulkhead into the center cabin, where many of the passengers were now gathered to watch the movie. They were too absorbed to notice him. He passed another bulkhead near the center of the aircraft and entered the rear cabin, populated by passengers uninterested in the movie. There was a couple to his right, a striking blond woman and a bearded man, holding hands, engaged in a quiet discussion. In the center tier was a mother and a silent, well-behaved boy, perhaps two years old. Farther back was another mother, traveling with two young children. She was seated, talking with a gentleman in the aisle—a large man with a thick, bushy red beard caressing a tiny baby held against his chest in a harness. A few rows behind them was another man, even larger, lost in the music from his earphones, but looking sullen.

Near the rear of the airplane, in the next-to-last row on his right, was a woman and a girl, not quite a teenager. The youngster was chewing gum, engrossed in a comic book.

Next to them, in the center row of seats, was flight attendant Annie Swift, a notepad on her lap. She glanced up as Desjardins approached and noted the no-nonsense expression on his face.

Swift rose and Desjardins grabbed her arm. "Get into the galley," he said.

When flight attendant Danielle Riendeau, standing only a few paces back, heard those words uttered in a tone of urgency and trepidation, her body stiffened involuntarily. Her eyes filled with spontaneous tears. Realizing that his tone had been a mistake, Desjardins placed a consoling hand upon Riendeau's shoulder. "Don't worry, Danielle," he said, "it's not serious."

Riendeau swallowed hard, took a deep breath, and willed the tears to go away. As she struggled to compose herself, she glanced at Annie Swift and immediately turned away. Swift's huge, expressive eyes silently asked: What's going on here?

The flight attendants—Annie Swift, Danielle Riendeau, Susan Jewett, Nicole Villeneuve, and Claire Morency—assembled about Desjardins, their eyes full of questions.

Desjardins opened his emergency manual. "Don't worry," he repeated softly.

7

The Cabin

"This is the nicest flight we've ever taken," Joanne Howitt said to her husband Bob, sipping an after-dinner glass of Drambuie.

Bob agreed, his outlandishly bushy, copper-colored beard and shoulder-length hair bobbing as he nodded his assent.

Dinner had been very pleasant, the smooth quiet ride relaxing. Three-year-old Brodie played contentedly with his collection of toy cars. Now, if Joanne could settle the baby, she could bask in a Drambuie glow and before she knew it arrive home in Edmonton. She was eager to call a halt to her maternity leave and begin the new job she had accepted.

"I'm going to change Katie," she said. She unbuckled her seatbelt, rose, cradled the newborn in her arms, and made her way through the airplane to a lavatory in the rear, next to the galley. She stepped past a couple of flight attendants, and entered the tiny cubicle.

Once inside, as she changed Katie's diaper, Joanne suddenly heard voices, muted, distant, crackling over some sort of intercom system. It's strange that they have an intercom in here, she thought. Perhaps it was a malfunction, an aberration in the electrical system. She cocked her ear to listen. "Winnipeg," she heard someone say. "We're going to Winnipeg," a voice said. Or was it, "We're not going to Winnipeg"?

Maybe it's the Drambuie, Joanne thought. Now I'm hearing voices.

Puzzled, but not concerned, she focused her attention upon Katie, who was growing restless. Saying a silent thanks to the inventor of disposable diapers, Joanne finished the task at hand, picked the baby up, resting her against one shoulder, unlocked the lavatory door with her free hand, stepped back into the cabin, and walked forward to her seat.

"Katie's getting cranky," she said to her husband.

"I'll take her for a walk," he offered. He reached for his daughter and lowered her into her Snugli, a harness that held her close to her daddy's chest. Then he rose and stepped into the aisle, an incongruous figure—a hulk of a father caring gently for his baby daughter.

He walked forward with no destination in mind, past a brooding man in the center rank of seats, absorbed in the music coming through his earphones, nursing a drink, attempting to lose himself in a copy of *Maclean's*, the Canadian news and variety magazine, but fidgeting, as though uncomfortable.

Bryce Bell had experienced better days. The six-foot three-inch, 215-pound former defensive tackle at St. Xavier University in Antigonish, Nova Scotia, was still in superb physical condition despite the passing years. Daily fifteen-mile jogs and a severe regimen of weightlifting had honed his body into even better shape than it had been fourteen years earlier, when he had been drafted to try out for the Winnipeg Blue Bombers of the Canadian Football League.

Bell had been impressive enough on the college football field to receive a modest $1,000 signing bonus from the

Blue Bombers, but he had journeyed to training camp with faint hope. He felt he lacked the speed and agility to play professionally. A week at camp had proved his self-assessment. He was cut from the team and cast out into the real world, where the necessities of life and the vagaries of fortune had trapped him in the routine of a stifling government job. Now, as director of program services in the Advanced Education Department of the Province of Alberta, Bell wondered how he had managed to get so far away from his dreams. He hated his job.

Bell had spent the past two weeks on holiday with his wife Margo and their only child, Jonathan, who was approaching his third birthday. Margo's parents owned a cottage forty miles north of Ottawa and Bell had hoped to enjoy a respite from Edmonton, from city life, from his job. But even the bucolic atmosphere of the lakeside cottage had not soothed this troubled beast. He had snapped at his mother-in-law, argued with Margo and, in general, behaved boorishly. He found it difficult to relax with the specter of his job haunting him. Two days from now, back in Edmonton, he would embark upon yet another unsatisfying forty-hour week of pushing student loans and grants. At least Margo and Jonathan would be spared his ill temper. They had remained behind to enjoy a few more days at the cottage.

To add to his snarly mood, Bell found himself on this balmy Saturday evening, when he would have much preferred to keep his feet on solid ground, sandwiched into the carcass of this aluminum bird, Air Canada's pride and joy, the 767. Although normally preferring a window seat, for this flight Bell had chosen a seat in the mid-section of the plane, behind the wings, in a non-smoking area.

He had flown on the 767 on his last three business trips and was not impressed. "The thing must have been wired by an idiot. You flick the switch on your overhead light and it lights up six seats away," he muttered to no one in particular. "Air Canada charges you $2.50 for earphones that don't work, and the bloody scheduling problems on 'the People's Airline' jerk you around." His most recent flight had been a fiasco, lowlighted by a four-hour delay as mechanics tinkered with the air conditioning. They could have this oversized behemoth, he had said to himself as he boarded, wondering how long the delay would be *this* time, for there surely would be one. Bigger was not always better, in the world according to Bryce Bell.

It was not merely the 767 or Air Canada that disturbed him. He was not one to take the miracle of flight lightly. His spirits were low and his anxiety high whenever he was obliged to fly. Although he was loathe to admit it, this powerhouse of a man was intimidated by airplanes. Religion was also low on Bell's list of priorities; nevertheless, he offered a small prayer of thanks each time the wheels of a 767, or any aircraft carrying his person, touched safely down.

All in all, Bell had spent the first hours of the trip concentrating upon his objective, which he would enunciate as "get the damn flight over with." He had reluctantly shelled out the $2.50 for earphones and immediately, to his grudging satisfaction, found Willie Nelson's voice rasping into his ears. Here was respite for his troubled soul. He and Willie would cruise at 41,000 feet and mellow out together.

Bell soon found that he had to sit extremely still to prevent static from interfering with the music. Worse, throughout the flight he was unable to suppress disquieting thoughts of the Air Canada disaster one month earlier. A Douglas DC-9 had caught fire during flight. The pilots had managed an emergency landing in Cincinnati, but had lost twenty-three passengers who perished from smoke inhalation before they could evacuate. The story was on everyone's mind these days, particularly Bell's when he was flying Air Canada. He read in the newspaper that part of the carnage had been blamed on the alcohol consumption of the passengers. According to one source, too many had their faculties impaired at a critical time, and had been unable to escape the flames. Perhaps, Bell mused, he should limit his intake of his favorite cocktail, one aptly named to suit his current mood, a Rusty Nail.

Smithfield ham—size arms popped out of his knit sport shirt as he adjusted his seat belt snugly across his trim midsection. He shifted his weight first from one side, then to the other, unable to find comfort. All the jogging, weightlifting, and exercise in the world had not prevented and could not cure his current torment.

Bryce Bell's hemorrhoids were killing him.

Dinner had been surprisingly good, even by Bell's demanding standards, and he had counted on the in-flight movie to take him the rest of the way. He was almost beginning to relax when he saw the in-charge flight attendant fiddling with the projector, and realized that the system was not working properly. His mood plummeted once more. Typical of Air Canada, he thought. A pretty blond flight attendant came around, apologized for the inconvenience, and told Bell he could move to a forward section of the airplane if he wished to see the movie. Gruffly he declined.

Foregoing his earlier resolution, he ordered another Rusty Nail, turned up the volume on his headset, picked up his copy of *Maclean's*, shifted his aching rump, and tried to get comfortable.

Some moments later he was annoyed when the music was interrupted by spasms of static, followed by the sound of a male voice over the intercom system. He put down his magazine and listened to the garbled transmission. The pilots were making an announcement, but it was not the usual weather report and "enjoy the miracle of flight . . .

thanks for choosing Air Canada" drivel that he routinely ignored. This announcement had substance to it.

There was some problem with fuel gauges, necessitating an unscheduled landing in Winnipeg to check things out . . . all passengers should return to their assigned seats.

Oh, for God's sake! Bell thought. Why can't they just fly the stupid airplane and get it there?

On days like this—days when she had to fly—Pat Mohr found herself edgy. At home that morning in Richmond, Ontario, a little too far southwest of Ottawa to be called a suburb, she had been in no mood to handle the complaint of her eleven-year-old daughter Heather, who whined, "Why do I have to wear what *she* wears?"

The Mohr family was embarking upon a three-week vacation with Ken Mohr's parents in Josephsburg, Alberta, a hamlet on the outskirts of Edmonton. For the trip Pat Mohr had purchased matching outfits for Heather and four-year-old Crystal, bright green T-shirts and similar green-and-white striped skirts, and she was determined that they would wear them. "Just get dressed, Heather," she said.

"I don't know why we have to leave so early," Ken Mohr had chided his wife gently. Whenever he flew alone on business, Ken was content to arrive at the airport with only time enough to hop onto the airplane before the doors closed; Pat insisted on getting there early.

"You know how I am," she had said, smiling. "I just feel better if we don't cut it so close."

Pat's job as a convention coordinator for the Canadian Nurses Association and Ken's position as manager of product engineering for Siltronics, Ltd., made them both frequent air travelers. Flying thrilled Ken, and he had even toyed with the idea of obtaining a private pilot's license. Pat, on the other hand, was less enthusiastic.

66

"Flying is *not* my most favorite pastime. I have that help-less feeling of not being in control," she confided to her husband.

They had arrived at the airport early and Pat had plenty of time in the passenger lounge for her anxiety to build. But as they walked out onto the tarmac to board their flight, she was pleased to realize they would be traveling on a 767. There was something about the size and the newness of the craft that she found reassuring. Ken was surprised that the Ottawa airport did not have a motorized ramp to fit the new aircraft. They had to step outside and climb an old-fashioned set of rolling stairs. The noise of the ground-based auxiliary power unit and kerosene fumes from taxiing aircraft had filled his senses.

Once inside, the spaciousness of the cabin was another pleasant surprise. "This is nice," Ken said to his wife, "You don't feel like you're jammed into a cigar tube."

"Yes, it is," Pat had agreed. "This is the largest plane I've ever flown on."

When they reached their assigned seats in the non-smok-ing section of the rear cabin, sitting four across in the center rank, Pat had said, "You know, it would be nice if both girls could have a window seat."

"Well," Ken suggested, "there don't seem to be too many people. . . . I'm sure we can move."

Pat hesitated. She worried that changing from their as-signed seats might tamper with fate. She preferred to stay where they were until they were airborne and the seatbelt sign was turned off.

Before takeoff Ken had pointed out the emergency exits to his wife and helped his youngest daughter fasten her seat belt while Pat did the same for Heather.

"Is there going to be a movie, Daddy?" Crystal wanted to know.

"Yes, there is," her father said. "But it won't be for a while yet."

The sisters quickly settled to their own diversions. Crystal always traveled with her personal bag of goodies, crammed with stuffed animals, drawing supplies, string beads, and other activities. Heather buried her nose in a comic book, a special first edition of *The Warlords*. For an eleven-year-old, she was an experienced traveler.

As Flight 143 taxied down the runway, Pat recalled a time not long ago when she had been flying with Heather and the cabin pressure had caused her daughter a painful earache. "Heather," she said, "chew some gum, just in case."

"Okay, what kind is it?"

"Bubble, of course."

Heather smiled at her mother. It was her favorite.

Heather stuck a pink wad of gum into her mouth and turned her attention back to *The Warlords*.

Only after the airplane picked up speed, jumped off the runway, and soared into the sunlit Canadian sky had Pat breathed a sigh of relief. Takeoffs and landings were the rough part, she felt. Once they were safely cruising at altitude, she could relax.

Finally, as the *ping* of the seatbelt signal granted permission, Pat and Ken Mohr decided to move so that both of their daughters could enjoy a window view. Pat and Heather went to the left side of the aircraft, in the next-to-last row. Ken took Crystal to the opposite side of the same row.

Almost immediately flight attendants had bustled about, offering juice, presenting each girl with a small winged Air Canada insignia. Crystal lobbied for an immediate dinner, and Pat and Ken shared a grin across the aisles. It was a family joke that Crystal only got twenty miles to the cookie.

When dinner arrived, Pat found her steak extremely pink and tough. She tasted it, and found it cold as well. "Heather, is your food cold?" she asked.

"Yeah, cold and gristly and raw!" Heather whispered, wrinkling her nose and making a face.

The heater trays must not be working properly, Pat surmised. She tried to catch her husband's eye across the wide body of the airplane but his head was turned as he helped Crystal with her food. She considered complaining, but decided against it. They had eaten before the flight and neither she nor Heather was particularly hungry.

After the dinner trays were removed Ken showed Crystal how to manage her earphones. They watched as a flight attendant punched a series of buttons, and they waited for Richard Pryor to cavort through *The Toy*, but the screen at the front of their cabin remained blank. A flight attendant apologized. "You are welcome to move to the front section if you'd like," she added.

Wouldn't you know? Ken Mohr thought with a bit of resigned irony. His job at Siltronics, Ltd., was to oversee the production of integrated circuits. "I may be a dinosaur, but I really don't trust this high-tech world," he told his friends. "I am right in the heart of this semiconductor business—making computer chips—and I see all the things that go wrong with these bloody things. There are a million and one things that can go wrong. Microprocessors work well and give you better service than mechanical stuff, but when an electronic system dies, there is nothing you can do with it. You can't tinker with it and make it run."

He was disappointed that the in-flight movie was one he had seen, but he knew that it would help to pass the time for Crystal, so he and his younger daughter moved forward, leaving Pat and Heather in the rear of the airplane. Crystal sat in an aisle seat, watching happily as Richard Pryor dizzily rolled around in an inner tube in a toy store.

Ken had been more interested in the conversation in front of him between a flight attendant and a passenger. He overheard flight attendant Susan Jewett offer to take

Nigel Field on a tour of the cockpit, and he was envious. This guy must be an Air Canada employee, he reasoned. I sure would like to do that.

The young woman left and moved forward in the direction of the cockpit and Ken returned to the business at hand, killing time. He did not notice Jewett return to tell Field that they were diverting to Winnipeg because of a mechanical problem. Rather, his first hint of trouble was a buzz in his earphones, cutting off the movie soundtrack. This was followed by a male voice that said: "Ladies and gentlemen, I would like to request that you return to your seats at this time. . . ."

Crystal was upset. "Daddy, why do we have to stop watching the movie?" she asked.

Sitting near the very back of the airplane, Pat Mohr heard the same announcement and thought, Oh, for crying out loud! What next? She glanced up to see Ken and Crystal returning to their seats on the opposite side of the cabin. What a shame, she thought. The whole projection system must be kaput. She and Ken exchanged shrugs of bewilderment and disappointment.

Everyone seemed to be milling around, vaguely uneasy. Everyone, that is, except Heather, who was still engrossed in *The Warlords*. Finally noticing the activity, Heather regarded the scene with detached curiosity. "Did somebody say we're going to Winnipeg?" she asked.

"Yes, honey, they have to fix something on the plane," Pat replied, attempting to squelch within her a rising sense of cold apprehension.

Bob Howitt, with baby Katie still snuggled into her harness against his chest, paced slowly through the right aisle of the aircraft. The vision of the hairy, bulky father and the tiny child brought amused giggles from two grandmotherly women seated on the right aisle next to the window. Howitt smiled and said hello.

They were sisters, Lillian Fournier and Pearl Dayment, leaving on vacation. "We're going to the mountains," Fournier announced. "Pearl's never seen the mountains."

Howitt learned that the sisters were from Pembroke, a logging town west of Ottawa, and he told them he had driven through Pembroke that morning on the way to the airport.

Dayment, seated next to the window, could not seem to tear her eyes away from the panorama below. Howitt looked out, too, and his educated eye made a quick geologic survey. As a soil scientist for the Alberta Research Council, he participated frequently in aerial surveys, although generally in light aircraft at a much lower altitude. Here, 41,000 feet over Red Lake, Ontario, he saw the unmistakable evidence of what is known as the Canadian Shield, rough topography interlaced with numerous lakes. These were igneous rocks, formed by the fiery cataclysms of volcanic eruptions and the upheavals caused by plate movements deep within the earth. The hard, rocky terrain was so different here than around Edmonton, where the surface was composed of softer limestone, sandstone, and shale.

Continuing his lazy stroll, Howitt moved forward and encountered a young mother, dressed in a dark skirt and white sweater, traveling alone with two small children. "That's a good idea," she said, pointing to Katie's harness.

"She likes it," Howitt replied.

Howitt and the young woman spoke briefly, comparing notes on the rigors of traveling with children, when they felt the unmistakable sensation of falling. For some strange reason the pilots had cut back power to the engines, as though they were beginning the descent to the airport.

Howitt knew that it was far too early to begin the approach to Edmonton. "I wonder why we've gone into a descending pattern," he said.

Moments later he knew why. There was some sort of problem with the fuel gauges. They were diverting to Winnipeg for an unscheduled landing.

Who gives a damn about the fuel gauges? Michel Dorais asked himself. Who cares? This is stupid. Why don't they just go on to Edmonton and fix the gauges there? He glanced to his left, to Shauna Ohe, sitting by the window. She was nervous, having been near panic since she felt the plane dip a few minutes earlier.

Dorais suddenly remembered with irony that they should not even be aboard this flight. They had planned to fly on Canadian Pacific Air, not Air Canada. "I'd rather give my money to private industry than to the government-owned airline," this champion of free enterprise proclaimed. But, weeks earlier, driving in downtown Edmonton, he had been unable to find a parking space in front of the CP Air ticket office, so he had driven on to Air Canada instead. "If we'd been on CP Air, we'd be home right now," he said to Ohe. "The only damn reason we're on Air Canada is that I couldn't find a parking space."

Things happen in threes, Richard Elaschuk thought as he led his two-year-old son Stephen down the aisle of the aircraft, back toward the seats on the left side where his wife Pauline sat, nursing baby Matthew, only four months old. Richard had taken his son forward to see the movie, and now he pondered the possible reasons for the sudden directive to return to their seats. Based upon his personal flying history, he had reason to be concerned. He had flown perhaps a half dozen times in recent years, and two of those trips had resulted in emergency landings.

A genial, broad-shouldered man of twenty-nine, Elas-

72

chuk was a product of Ukrainian stock, whose great joys in life were his wife and sons, and the sport of curling, which, in parts of Canada, takes the place of bowling as the common man's recreation. A Scottish sport, curling is played in teams on an ice rink that resembles a bowling alley. One person slides a curling rock, fashioned of special stone, toward the target area. Teammates guide its progress by sweeping ice away from, or toward, the stone, as the situation demands.

Years earlier, Richard had needed a fourth curler for a *bonspiel*, a curling tournament. He had asked Pauline and she, knowing nothing about the sport, had accepted. That first date led to marriage, family, and a modern new home in Westlock, Alberta, about an hour's drive north of Edmonton. The years that followed took them on occasional trips to the east, for Pauline hailed from Prince Edward Island on the opposite side of the country. In five years of marriage, they had traveled "down east" several times, although they did not always fly together. Their job schedules at Immaculata Hospital in Westlock, where Richard was a pharmacist and Pauline a nurse, sometimes necessitated separate itineraries.

Twice now these trips had put Richard's life in jeopardy.

On the first occasion, he was on a flight from Edmonton to Winnipeg, to link up with Pauline on Prince Edward Island. He had already realized that the aircraft was taking too long to land, when the first officer emerged from the cockpit and, standing in the aisle, announced, "We're having a possible problem with the landing gear. The panel indicator in the cockpit shows that it is not locked in the down position. It could just mean that the light is out, but I'm going to look at it."

To Richard's amazement, the first officer then rolled up the aisle carpeting to expose a door. Opening the hatch, he leaned well down into the landing gear bay, contorting his body until he was hanging upside down, and inspected the assembly with a flashlight. He remained in this awkward position for several seconds before easing back up into the passenger cabin. He closed the hatch, replaced the carpeting, announced tersely, "Everything is fine," and disappeared back into the cockpit.

Minutes later, when the aircraft landed in Winnipeg, the flight was met with fire trucks and ambulances aligned along the length of the runway, their flashing red lights a clear, garish message that the situation had been more critical than they had been told. The incident had left him shaken. He realized how vulnerable and helpless he was on any airplane.

On a second trip Richard had been sitting at the rear of an Air Canada DC-9, back near the noise of the twin tail-mounted engines, when he realized that one of them was malfunctioning. The steady loud drone was replaced by a peculiar fluttering noise. The pilot had announced that the aircraft was diverting to Montreal because of "minor technical problems." "What the hell is going on?" a passenger in front of Richard had demanded of a flight attendant, but he had received only a noncommittal reply. Once more the aircraft landed amid a complex of emergency vehicles, and Richard, telephoning Pauline, learned that the flight's engine problems and emergency landing had been covered by the national news.

By July 23, 1983, therefore, Richard had ample reason to suspect any indication of "minor problems" on a commercial flight. His eyes met Pauline's as he walked past to strap young Stephen into his window seat. He took his own seat, next to Stephen and behind Pauline and baby Matthew, and he thought, We're in trouble now. They want us back in our assigned seats in order to balance the plane. Then another, more terrifying, reason leapt to mind: I guess they also want to know whose body it is when we hit the ground.

74

8

The Cockpit

Using a frequency reserved for intra-company transmissions, Quintal radioed the Air Canada office at Winnipeg International Airport to report the imminent and unexpected arrival of Flight 143. Briefly he explained that they were experiencing difficulties with the fuel feed system. Mechanics at Winnipeg were placed on alert, ready to check the aircraft thoroughly when—and if—it arrived.

Once more the three men in the cockpit of Flight 143 heard the persistent, unwelcome sound of a series of four warning beeps. In front of the captain and first officer, the EICAS screen continued to proclaim the dismal news: All six fuel pumps were failing. Pearson muttered, "I just hope this is . . . false warnings." Then he asked Dion, "Can you think of anything we haven't done?"

"No, I can't, Bob."

Four more beeps sounded, followed by a few moments of stark silence, then four more beeps. Pearson had never been in a situation so difficult to predict. Clearly the engines were starved, but there was no way to assess how much time they had before the pumps gave out altogether, before they stopped feeding fuel to the engines, before the engines quit. There were no direct-reading fuel gauges to provide answers to these questions, and no one in the cockpit retained any faith in the numbers provided by the flight management computer.

Pearson maneuvered the aircraft gently, cautiously, so that he would not have to demand an iota of extra performance from the engines. He queried Dion on his opinion of the best method to maintain fuel supply to the engines. Pearson made an educated guess: If they kept the aircraft in a flat attitude, rather than nose down, the engines would be better able to suck in whatever fuel remained in the lines. Dion agreed.

Four more insistent beeps sounded, disrupting conversation. The cockpit was assailed by eerie warning buzzers and foreboding color-coded lights alerting them to a series of problems escalating in severity.

Nine seemingly endless minutes had passed since the onset of the crisis when a single, sharp *bong*! jolted the three men. It was a sound they expected, but dreaded.

"Okay, we've lost the left engine," Pearson confirmed.

The two pilots commenced an immediate drill. They had lost an engine. Certain things must be done.

"Power and gear as required," Quintal snapped.

"Check," Pearson replied.

"Throttle closed, auto throttle disengage."

"Disengage."

"I'll disengage now."

"Okay, we are . . ."

Pearson's words were cut off by four warning beeps. Right engine oil pressure: *falling*. Right engine temperature: *low*.

Quintal glanced at Pearson, who appeared calm at the controls, then he lectured himself: I'm not going to help this guy by being nervous. If I am a nervous wreck I will not be any help; I will be a nuisance. My job is to think, "What is he going to need now?" Before he asks, I'm gonna give it to him. Quintal willed his hands to stop shaking. I must start talking, right? If I control my voice, everything will be fine.

He was relieved to hear professional coolness in his voice as he apprised Winnipeg ATC of the critical events. "We've lost our number one engine. We'll require all the trucks out."

9

Winnipeg ATC

"One forty-three, check, okay," Ron Hewett acknowledged into his microphone at the Winnipeg ATC Center.

Arrival controller Len Daczko, hearing the exchange over the loudspeaker that flooded the room with Quintal's voice, picked up a hotline connected directly to the airport control tower. He advised the tower to have the emergency equipment ready on the runway.

Steve Denike, the administrative supervisor, alerted the Accident Investigation Board that a major incident was under way. Whatever the outcome, every minuscule detail of this incident would by law be examined and reexamined.

Within seconds, the controllers heard Pearson's voice requesting clearance to runway 18 instead of the originally assigned runway 31. It was the first time the captain had spoken to them directly, and it was clear evidence of a deteriorating situation. The pilot had changed his strategy. Pearson was going to set the aircraft down without wasting a single moment.

Runway 31 ran from the southeast to the northwest; to

reach it, Pearson would have to circle to the south, using precious time and fuel. Runway 18, on the other hand, ran directly north to south and was straight in the path of Flight 143. There was another factor. Eighteen was the longest runway at the airport.

Hewett double-checked the winds; they were too light to affect a landing on either runway. He cleared Pearson's request with Daczko, who quickly agreed. The controllers knew that the cockpit was now an environment of controlled frenzy. Pearson and Quintal had their hands full. For any major development such as the loss of an engine, they immediately had to complete the procedures called for in a memorized drill. They then had to follow the directives of a printed checklist. On top of that they had to navigate their way to Winnipeg, calculate and monitor the correct descent profile and prepare for landing. The primary job of the controllers was to keep the airspace clear, and this was a relatively easy job amid the light traffic of a Saturday evening. Beyond that, they stood ready to assist, but it was not their place to offer alternatives that would confuse the issue. The pilots, riding with the aircraft, were in the best position to make the critical calls. For the most part, therefore, Hewett would not initiate transmissions.

But now that the pilots had requested emergency equipment on the runway, regulations forced Hewett to ask for certain information. In the event of a crash on the runway, rescue teams had to know how many people were involved and firefighters had to gauge the potential of a catastrophic explosion.

So Hewett requested: "One forty-three, when you've got the time, we'd like your fuel load on landing, and the total number of persons on board."

10

The Cockpit

Quintal was too busy to deal with the controller's question, so he ignored it. Having completed the memorized drill, he and Pearson systematically reviewed the single engine landing checklist. "Approach and landing, flaps will be twenty," he said.

"Right," Pearson confirmed.

Flaps, huge panels on the trailing edges of the wing, are normally lowered to their maximum angle of thirty degrees for landing. The effect is dramatic, decreasing the velocity of air beneath the wings, piling air particles on the underside, increasing the lift. Flaps enable a jetliner to fly at considerably lower speed, producing a safe, controlled descent and landing. At a normal flap setting of thirty degrees, a 767 lands at a speed of 116 knots to 153 knots, depending upon aircraft weight.

With only one engine operating, however, the checklist instructed the pilots to land with the flaps set at only twenty degrees. This was a tradeoff. The lesser flap setting would provide the pilots with greater control of the aircraft during the final portion of the landing run, compensating somewhat for the lost engine. However, it would produce a landing speed some ten knots greater than normal. If the aircraft encountered any trouble on the ground, the greater speed could magnify its effects.

The sophisticated computer bank of the 767 did not cope well with deviations from normal operating procedure. As a safety precaution, the computers would automatically alert the pilots with loud oral warnings if they failed to set the flaps at thirty degrees. To prevent this distraction, the pilots had to deactivate the automated system.

"Ground proximity flap override switch-override," Quintal intoned, reading off the checklist. "That's the one there—override."

"Okay, ground flap override. All right, go ahead," Pearson said.

"Final approach speed will be twenty flaps, speed plus the wind. Stand by for . . . it says we weigh 98,000. . . ."

Based upon the 98,000 kilogram weight of the aircraft, as reported by the flight management computer, Pearson and Quintal set their "bugs," movable white tabs affixed to their primary airspeed indicators, as a visual reference to the proper approach and landing speeds. There were four bugs to delineate the minimum safe speeds at each of the four flap settings they would use during approach: 180 knots was the minimum safe speed before the flaps were lowered, 139 knots with leading edge slats, 133 knots with five degrees of flap, and 129 knots with twenty degrees of flap at touchdown.

By now Flight 143 had descended to 28,000 feet in its approach to Winnipeg. Suddenly, at twenty-two seconds past 0121 GMT, the cockpit was plunged into darkness. The bright, color-coded, easy-to-read data units provided by the flight management computer, the bank of digital displays that reported airspeed, altitude, compass direction, navigational data, engine speed, temperature and RPMs, fuel flow, oil quantity, pressure and temperature, even the

82

clock and the three temperature gauges—the entire array of "gee-whiz" electronic gadgetry in the cockpit of the world's most sophisticated airliner—vanished in an instant.

The glow of the late afternoon sun illuminated the faces of the three men in ghastly relief.

"How come I have no instruments?" Pearson asked, incredulous.

The answer was as simple as it was terrifying. The space-age technology of the 767 cockpit feeds upon electricity supplied by generators run by the two massive engines. The engines, in turn, are powered by Type A-1 jet fuel. It had never happened before—in fact, neither Boeing, nor Air Canada, nor Pearson, nor Quintal, nor Dion had ever contemplated the scenario—but if a 767 runs out of fuel, a diabolical domino effect takes place. The engines quit. In turn, this stops the generators, halts the production of electricity, and transforms the computerized cockpit displays into darkened, totally useless, cathode ray tubes (CRTs). To Pearson it seemed as if the cockpit had become the darkest place in the world.

This is what is happening, Quintal thought. But another part of his brain argued: It is impossible. This doesn't happen to me.

But the unthinkable *had* happened. Captain Bob Pearson and First Officer Maurice Quintal now found themselves 28,000 feet over central Canada in a Boeing 767 loaded with sixty-one passengers and eight crew members, still more than 100 miles from Winnipeg, with less instrumentation and fewer controls than a Piper Cub. They now faced the consequences of an extraordinary series of malfunctions, blunders, and unfortunate coincidences.

Incredible as it seemed, they had run out of fuel.

Pearson and Quintal consulted yet another emergency checklist, detailing the procedures to follow in the event of a total loss of power. The first task was to start the Auxiliary Power Unit (APU), which would provide electrical, hydraulic, and pneumatic support. Pearson turned and asked mechanic Rick Dion, "What do you think?"

"You've got nothing to lose," Dion reasoned. "But I don't think it will run too long." Dion came to that conclusion because he knew the APU ran on the available fuel supply.

Quintal threw the switch. A few instruments flickered to life.

"Something came back," Quintal said.

Dion suggested that they once again open the cross-feed valves to the center fuel tank, holding out a slim hope that some kerosene remained there from a previous flight. Pearson tried this. Nothing. There was no question that they were out of fuel.

Quintal flipped hurriedly through the Quick Reference Handbook until he found instructions for unlocking the ram air turbine (RAT), dropping it from its position near the right wheel well. Housed in a nacelle of its own, like the far larger but now useless jet engines, the RAT utilized wind power to provide a minimal level of hydraulic control. Air, scooped in through its nose, turned a four-foot propeller that provided sufficient power to enable Pearson, with effort, to manipulate the ailerons, elevators, and rudder, but not to lower the flaps and slats, move the stabilizer, activate the speed brakes, lower the undercarriage, or provide normal braking and reverse thrust upon landing.

Pearson tested its effectiveness by manipulating the stick and the rudder pedals. He recalled an incident when he was driving a car and the engine quit, rendering the power steering useless. To control the car, he had to manhandle the steering wheel. This was similar. The single most important consideration was to remain in *control*.

"Winnipeg, AC 143," Quintal radioed.

"AC 143 go ahead," Hewett replied.

Quintal said, "Just lost both engines."

11

Winnipeg ATC

He's gliding, controller Ron Hewett said to himself. He's in a great deal of trouble.

Hewett was a veteran. In the old days an airliner would occasionally experience icing problems in the fuel lines; once in a while there was a problem with contaminated fuel. But in the modern world it was rare—almost unthinkable—for a jetliner to lose all power. Whenever he had to deal with that kind of problem today, it usually involved light aircraft experiencing difficulty because of a mechanical problem in a lone engine, or perhaps because the pilot was lost. He had not dreamed of encountering this problem in a 767.

Even as he heard the improbable, inexplicable words of Quintal, Hewett studied his radar screen intently. The blip that had signaled Flight 143's presence and the computerized triangle of flight data suddenly disappeared.

Flight 143 seemed to have vanished from the sky.

12

Flight 155

"That's Maurice," First Officer Gilles Sergerie said to his captain. "I want to hear this."

Sergerie was the first officer on Air Canada Flight 155 from Montreal to Calgary. His Boeing 727, trailing Flight 143 by about half an hour, followed a similar airway to the west, albeit at a lower altitude. He, too, was being guided by Winnipeg ATC, and was monitoring the same radio frequency as Flight 143. He would have paid close attention to the unfolding drama in any event, but Sergerie had a personal interest. The voice he heard relaying the distress signal was that of his close friend and neighbor, Maurice Quintal.

"That's too bad," Sergerie had said when he had heard that Flight 143 had lost one engine.

The situation at that point was far from routine, but, in the professional judgment of the airborne eavesdropper, it was not critical. An aircraft can lose an engine for a variety of reasons and still fly safely. Sergerie had not heard the one word that would trouble him most, the word that was on everyone's mind since the Cincinnati disaster a month earlier—fire. If there was no fire, chances were good for a successful emergency landing.

Sergerie had gone about his business, a bit more tense than usual, but optimistic. He heard controller Hewett's query concerning the number of passengers on board and the estimated fuel load upon landing. Quintal had ignored the question at first but now, a few minutes later, Sergerie was shocked to hear Quintal report, "I am out of fuel, ah, come back with the ah, number of passengers . . ."

"Wow!" Sergerie exclaimed.

He lit a cigarette in an instinctive reaction. Streams of exhaled smoke rose in the 727's cockpit. History recorded a few vaguely similar examples: a Republic Airlines DC-9 made a successful forced landing at Luke Air Force Base in Arizona with only five gallons of fuel remaining in the tanks; a Pan American World Airways Boeing 747 lost three of its four engines shortly after touchdown at Newark, New Jersey, because it was running out of fuel; a United Airlines Douglas DC-8, while awaiting final landing clearance, exhausted its fuel supply and crashed. Never before, however, had a commercial jetliner completely run out of fuel—without warning—while at cruising altitude in the very midst of flight.

"How could this happen?" Sergerie wondered aloud. "You don't just *run out* of fuel!"

13

The Computer

On shelf E2-4 of the main equipment center, located in the belly of the Boeing 767, sits one of the marvels of the electronics age. It is a digital, dual-channel computer known as the fuel quantity processor. Compact enough to be carried by hand, it is nevertheless a complex construction of microcircuitry, for its job is one of the most critical among all of the 767's computers. That task is to supply the cockpit crew with an up-to-the-second accounting of the fuel load. It accomplishes this by means of an intricate system of electronics networked throughout the aircraft.

That system begins in the fuel tanks, which are complex structures in their own right. Laymen tend to think of an aircraft's wings as simply large, hollow fuel tanks, but this is an impossible oversimplification. If such a large holding tank was fashioned as a single tube, it would prove disastrous. The enormous quantities of fuel carried inside would ebb and flow within the wings during flight, creating fatal instability. To protect against this, each of the three fuel tanks of the 767 is separated into compartments, fourteen per wing tank and four in the smaller center tank.

The compartments not only store fuel and assure stability, but house electronic sensors, wired to measure the volume, weight, and temperature of the fuel within. In an extraordinary leap of technology, Boeing engineers and subcontractors designed the 767 fuel tanks with the ability to report on their own contents. During each second of flight, a swarm of electronic messages flashes between the fuel tanks and the computer on shelf E2-4. Digesting this data, and adjusting for variables in temperature and altitude, the fuel quantity processor flashes a real-time fuel status report that appears in the cockpit on bright, easy-to-read liquid crystal displays.

Because of the essential nature of its data, the fuel quantity processor double-checks its own calculations on either of two independent and redundant operating channels. The two separate channels run off dedicated electrical power sources connected to one another through a device known as the "motherboard." Each channel is capable of performing the job without the aid of the other and the system is so sophisticated that the computer is able to monitor the quality of its own performance. Moment by moment, the fuel quantity processor determines which of its two channels is operating more efficiently and then selects for that channel. It will shut down a faulty channel altogether. To the designers of the system, the mechanics who served it, and the pilots who trusted their lives to it, the fuel quantity processor seemed almost fail-safe.

None of them, however, could peer deep within its circuitry. If they had the training, ability, opportunity, and equipment to do so, they would have found, in the fuel quantity processor of Flight 143, an aberration buried within the largest of six inductors encapsulated in an epoxy compound. Digging inside this inductor, they would have discovered a "cold-soldered" joint—a partial and improper connection—that controlled the operation of channel 2 of the fuel quantity processor.

Under normal operation, 5 volts of current traveled through this joint, causing channel 2 to operate properly. If the connection was completely bad, no current would have passed across and the fuel quantity processor, per its design, would have deactivated channel 2 and operated efficiently on channel 1. The partial connection, however, was a critical flaw, for it allowed about 0.7 volts to flow through, and this was just enough to default the default system. Because insufficient current flowed, channel 2 did not operate properly. But because *some* current flowed, the entire system was fouled. The computer was not designed to cope with a partial flow and, in effect, threw up its hands in despair and stopped working altogether.

This became apparent at Edmonton International Airport the night before the fateful journey of Flight 143. It began when thirty-one-year-old maintenance technician Conrad Yaremko was assigned to clear up a baffling problem on Aircraft 604 in an attempt to prepare it for a morning flight to Ottawa and Montreal and a return trip later in the day which would be designated Flight 143. Yaremko was informed that the fuel quantity processor was malfunctioning.

The aircraft in question was the forty-seventh production model 767 to roll off the Seattle assembly line of the Boeing Aircraft Corporation. In service for only three months, its fuel quantity processor had been suspect for much of that time. Intermittent problems over the past few weeks had indicated some lingering internal malfunction, some snafu buried deep within the intricate electronic network, but no one had been able to pinpoint it. The most obvious solution was to simply replace the microcomputer, but there was no spare fuel quantity processor available.

Yaremko knew he would have to make adequate repairs or the airplane would be grounded.

When Yaremko made his way to the cockpit of Aircraft 604, he found the fuel gauges to be blank. This was perplexing, because the two-channel system of electronics should have compensated for any problem. Even if one channel failed altogether, the processor should operate efficiently on the other, and Yaremko thought it unlikely that both channels had failed.

Following the procedures detailed in his maintenance manual, Yaremko searched for the source of the trouble. First he needed to assure himself that the processor was receiving its proper supply of electrical power. He located the circuit breaker panels in the cockpit, above and behind the captain's seat. He pulled the circuit breakers of both channels into the off position, waited about ten seconds, and then pushed them both into the on position. Then he left the cockpit and made his way to the main equipment center in the belly of the aircraft. On shelf E2-4 he located the fuel quantity processor and activated the Built-in Test Equipment (BITE) system, the feature that allowed the computer to monitor its own performance.

Selecting for channel 1, Yaremko depressed a button marked "Press to Test." A digital readout flashed the number 88.8, the coded signal that channel 1 was working properly. He then selected for channel 2 and repeated the test, but the digital readout remained blank. Clearly there was a fault in channel 2. Just as clearly, channel 1 should have taken over automatically, bringing the fuel gauges to life. Why had it not done so?

Like any other troubleshooting mechanic, the perplexed Yaremko decided to experiment. He returned to the cockpit and once more pulled off the circuit breakers for both channels. Then he reset only channel 1. Within seconds the fuel gauges glowed with life, their digital readouts indicating a total fuel load of 4,400 kilograms remaining from the previous flight.

Yaremko had done nothing to fix the mysterious problem deep within the fuel quantity processor, but he had found a way around it. By chance he had stumbled upon a way to circumvent the glitch. He had no way of knowing about the cold-soldered joint in channel 2, but he learned empirically that as long as the circuit breaker for channel 2 remained off, channel 1 appeared to work properly, providing all necessary fuel quantity information. To make sure no one countermanded his makeshift solution, Yaremko covered the channel 2 circuit breaker with yellow maintenance tape marked with the abbreviation "inop." The bright yellow tape served notice to all that the circuit breaker was to remain in the off position until further work was performed by maintenance personnel. This was a common temporary solution to electronic malfunctions.

Even with the fuel gauges working, Yaremko was not yet satisfied that the aircraft was airworthy. It was not his decision whether it was safe to fly with only one operational channel on the fuel quantity processor. This was a decision that had to be made by the book.

Airliners are built upon the principle of redundancy. A machine as complex as a jet airplane is not expected to perform with 100 percent of its components working. Parts fail; systems malfunction. Every vital component, therefore, has one or more combinations of systems and procedures that function as backups. The passenger is unaware of how often these redundancies are employed. The fact is, minor technical problems are the norm, not the exception, particularly on a new aircraft.

Certain systems are more vital than others. Some backups are adequate for the normal conduct of a safe flight; others are designed for emergency use only. To this end, Air Canada, as do all airlines, publishes a Minimum Equipment List (MEL), spelling out in detail the minimum equipment necessary for safe, routine flight. Yaremko checked the MEL to determine its dictates in the event only one channel of the fuel quantity processor was working. He noted in the logbook of Aircraft 604: "Fuel quantity indication blank. Channel 2 at fault. Fuel quantity 2 circuit breaker pulled and tagged. Drip required prior departure." The final comment indicated that, according to section 28-41:02 of the MEL, the airplane could be flown with only one operative fuel quantity processor channel, as long as a manual measurement was performed to verify the initial fuel load. Known as a "drip" or "dip," the procedure was as old as manned flight, familiar to any mechanic. It would simply and easily confirm the single-channel reading of the fuel quantity processor.

Prior to takeoff from Edmonton, mechanics performed the drip procedure and it did confirm the fuel load as reported by channel 1. Twice the fuel gauges had gone blank, but these incidents occurred on the ground, during servicing. The flight to Ottawa and Montreal was otherwise uneventful and, by midafternoon, Aircraft 604 stood at its gate in Montreal ready for servicing for the return trip.

Jean Ouellet relaxed in the smoke room provided for Air Canada hangar crews at Montreal's Dorval Airport. The forty-three-year-old mechanic, a dark-haired man of medium height, soft-spoken and painfully shy, was a seventeen-year veteran of the Air Canada maintenance force. Six weeks earlier he had completed a two-month training course that prepared him to work on the Boeing 767, but he had not yet been called upon to do so. He had returned the day before from a two-week vacation, and was savoring the final minutes of leisure before his shift started at 3:20 P.M., when he was approached by his foreman, Gary Geldart.

94

"You'll be working on a 767 tonight," Geldart informed Ouellet. "Six-oh-four is coming in with a problem. It needs a fuel drip. Bourbeau will be with you to give you a hand." Ouellet knew fifty-two-year-old Rodrigue Bourbeau only casually; they had worked together a few times in the past.

Ouellet gathered his tools, loaded them into his truck, and drove down to the maintenance office near the ramp where he studied the microfiche data that detailed the 767's fueling system and the procedures for verifying the fuel load manually by means of a drip. He located the pages with the appropriate data and printed hard copies.

Eight drip sticks are located at various positions underneath each wing of the 767. Each resembles the oil dip stick on an automobile, but it is mounted upside down. It serves as the axis of a magnetic float, called a doughnut, that sits within the fuel tank. Rotated counterclockwise, a drip stick releases from its housing and falls until the doughnut contacts the surface of the fuel. The depth of the fuel at that particular location is then indicated by calibrated marks on the drip stick.

Since the 767 wing is complex, backswept with a dihedral (an upward angle), readings must be taken at various locations. The mechanic then accounts for other variables. Airport ramps are built to slope away from the terminal to allow for drainage, and the fuel level at various points within the tank varies accordingly. The drip stick readings must be correlated with the forward-backward (pitch) and left-right (roll) angles of the aircraft. This information can be obtained from inclinometers inside the left main landing gear housing, or from another set of inclinometers in the cockpit. By comparing all this data, a mechanic can calculate with great accuracy the volume of the fuel on board. To the novice, the procedure and its attendant calculations may sound complex, but to mechanics such as

Ouellet and Bourbeau, it is part of the professional routine. Whenever there is any uncertainty concerning the fuel load it is generally confirmed by means of a drip.

As Ouellet refreshed his memory on the drip procedures, Bourbeau entered the room. "Are you checking for the drip on 604?" he asked.

"Yes," Ouellet replied.

"We're going to work together." Bourbeau scanned the instruction sheets Ouellet had printed. "How is it?"

"Well, it doesn't look too bad," Ouellet answered. "It looks quite straightforward."

"Would you print me a couple, too?"

"No problem," Ouellet said. He printed an extra copy of the instructions. Then he checked the time. It was about 3:50. "Well, we better go," he said. "I guess the airplane has arrived by now."

Neither Ouellet nor Bourbeau knew why the drip was required, but they knew that the aircraft's logbook would explain. This is what they would consult first. Each man drove a truck out to the gate where Aircraft 604 sat, awaiting service. The passengers had already debarked and, as the two mechanics reached the cockpit, the incoming pilots, Captain John Weir and First Officer Don Johnson, brushed past them on their way out, leaving them alone in the giant, empty aircraft.

Bourbeau grabbed the logbook from the pedestal in the center of the cockpit. He sat sideways in the first officer's seat and opened the journal on his lap. Ouellet stood next to him. They read Yaremko's note, written in Edmonton: "Fuel quantity indication blank. Channel 2 at fault. Fuel quantity 2 circuit breaker pulled and tagged. Drip required prior departure." This seemed peculiar to Ouellet, as it had to Yaremko the night before. If one channel was malfunctioning, the other channel was supposed to take over automatically. Why would there be any need to pull the circuit breaker? Neither Ouellet nor Bourbeau made a thorough check at this moment, because they were pressed for time. Had they done so, they would have realized that the fuel gauges were working, despite the ambiguity of Yaremko's note. But another 767 had landed and taxied to the adjacent Gate 4; they had to service that aircraft also. Bourbeau left to check on the other aircraft and Ouellet retreated to the passenger cabin to attend to another problem, a minor difficulty with the oven in the rear galley.

Ten minutes later, after repairing the galley oven, as Ouellet made his way back toward the cockpit he glanced out of a window on the left-hand side to see if the fuel truck was in position. It was not. This was likely to cause a delay, for the mechanics obviously could not complete the fuel drip until the fuel was on board. He thought to himself: Well, I've got a couple of minutes until the truck shows up. I'll go do a BITE test on the processor. He had not been instructed to do so, but he was professionally curious about this particular problem, and he decided to take the initiative. It was, after all, his first chance to familiarize himself with the much-ballyhooed new aircraft.

Ouellet knew that the first procedure was to recycle the circuit breakers. Reaching above him, he located the circuit breaker for channel 2 of the fuel quantity processor. Without removing the yellow maintenance tape applied by Yaremko—the tape clearly marked "inop"—Ouellet reset the circuit breaker to the on position, unknowingly activating the glitch caused by the cold-soldered joint. He then made his way down to the electrical bay to perform the BITE test. He tried to test channel 2 and found, in his words, "all kinds of weird things happening," so he switched to channel 1. He received a reading of 88.8, which told him that the channel was operational. He was as confused as Yaremko had been the night before in Edmonton. If channel 1 was operational, the fuel gauges should be working, but according to Yaremko's notation in the logbook, they were blank. Seeking further information, he returned to the maintenance office to study the microfiche, unaware that, by pushing on the channel 2 circuit breaker, he had caused the fuel gauges to go blank.

In the Flight Operations Office Captain John Weir, having just arrived from Edmonton and Ottawa on Aircraft 604, spoke briefly with First Officer Maurice Quintal. "Are you going out on Flight 143 with Aircraft 604?" Weir asked.

"Yes," Quintal replied. "How's the aircraft and the route?"

"There's a deviation on the flight plan," Weir reported. "Six-oh-four has a fuel problem." Weir said that, according to the MEL, this would necessitate a drip and lengthen the refueling process, so he suggested that Quintal might want to deviate from the normal procedure. Instead of boarding a safe minimum fuel load from Montreal to Ottawa and then taking on more fuel for the longer leg to Edmonton, Weir suggested that Quintal board enough fuel here in Montreal for the entire trip. That way Flight 143 would not experience a second delay by refueling in Ottawa.

Quintal agreed. He called Air Canada's Flight Dispatch Center in Toronto and requested a change in the flight plan, authorizing the extra fuel load in Montreal.

Wearing a gold-buttoned blue blazer with the winged emblem of Air Canada pinned to his left breast pocket, toting an overnight bag and a briefcase full of operating manuals, Captain Bob Pearson strode easily, confidently, across the employee parking lot toward the airport terminal. Halfway across the lot he encountered his good friends John Weir and Don Johnson. All three men flew 767s for Air Canada and also played hockey together. Weir was a neighbor of Pearson's in Beaconsfield, a western suburb of Montreal. Johnson and Pearson worked together for the Canadian Airline Pilots' Association (CALPA), Johnson as the representative of Montreal-based pilots and Pearson as a contract negotiator. The three men stopped for a few minutes to talk shop, giving Pearson his first indication about the glitch in Aircraft 604.

In one sense, this was not Pearson's problem, due to a new and controversial change in airline procedures. Boeing had designed the 767 to operate either with the full complement of three pilots (a captain, a first officer, and a second officer—also known as a flight engineer) that had become the standard of the industry on larger jetliners—or with only a captain and a first officer. Boeing left it up to the airline regulatory bodies to determine whether modern technology had rendered the flight engineer obsolete.

Initially Air Canada planned to utilize three pilots in the 767, but when, after lengthy and heated debate, the Federal Aviation Administration authorized U.S. airlines to operate the 767 with only two pilots, Air Canada followed suit. Members of CALPA agreed to the limitation, knowing that the future of their jobs depended upon the economic vitality of their government-owned airline. Air Canada could not pay the salary of a third pilot and remain competitive with U.S. airlines. "We didn't want to be involved with feather-bedding," Pearson reported. "We got assurances from the airline that the pilots would not be overburdened with work. They weren't going to take the work of three and slough it off on two."

The decision necessitated a rewrite of the operating manuals. Even with the assistance of the 767's wondrous computers, two pilots could not attend to all the details previously handled by three. By necessity, therefore, some of those duties were given over to ground personnel. One substantive change was to withdraw from the pilots the responsibility to oversee the fueling process. During their training, 767 pilots were told that this duty, previously handled by the flight engineer, was now the domain of maintenance personnel.

But in a larger sense, Pearson knew that everything was his responsibility. An airplane aloft, like a ship at sea, is a nation unto itself. To be sure, support systems abound, such as air traffic control, radar, and weather reports, as well as the entire on-board technological package. Under routine circumstances, a pilot will obey the commands of the air traffic controller and follow the rules of airborne navigation. Extraordinary occurrences negate all this. The simple fact is that an airliner, once in the sky, is an empire ruled by its captain. The buck stops there.

Thus, Pearson knew that, insofar as the fuel situation affected the safety of his charges, it *was* his responsibility. So he listened carefully as Weir pinpointed the problem on Number 604. None of the three pilots understood the technical implications, nor were they required to. What did concern Pearson was that the snag would delay departure by at least fifteen minutes, and he paid attention to Weir and Johnson's explanation of how they had saved time by fueling the plane only once, taking on enough fuel in Edmonton for the entire trip. "Sounds good," Pearson said. "We'll do the same thing."

Realizing that he could not fix the fuel quantity processor, mechanic Jean Ouellet said to himself: I might as well go up and pull the breaker and leave it as is. He returned to Aircraft 604 and approached the cockpit, only to find it crowded with service personnel performing their various functions. Rod Bourbeau had returned also.

Before Ouellet could enter the cockpit to deactivate the circuit breaker for channel 2, his attention was diverted. "We haven't got much time," Bourbeau said.

Ouellet agreed. They had to drip the tanks now to calculate the amount of fuel left over from the previous flight. Then the refueler would know how much to add. After that, a second drip would be required to confirm the total fuel load.

The press of the work load forced a fundamental error: Ouellet failed to deactivate the channel 2 circuit breaker. If he had done so, the fuel gauges would have come to life, as they had done for Yaremko. But as long as the circuit breaker remained on, the fuel gauges would continue to exhibit that peculiarly disconcerting nothingness produced by a recalcitrant computer.

"I've never done this on a 767," Bourbeau said.

Neither had Ouellet, but he had dripped many other types of airplanes, and he knew this would be a similar procedure.

The first job was to establish the attitude of the aircraft. Bourbeau positioned a stepladder and ascended into the left wheel well, where he noted that the inclinometers indicated that the aircraft wings were angled 0.5. The nose was level.

The mechanics knew there would be little fuel left in the tanks after the long flight from Edmonton and Ottawa, so there was no point in checking the drip sticks at the higher, outer edges of the wings. Consulting the microfiche pages, they located the number 2 drip stick on the underside of the left wing, close to the fuselage. They maneuvered a truck underneath the left wing into the small space between the fuselage and the engine, and then climbed up to perform the drip.

Ouellet showed Bourbeau how to twist the stick loose from its moorings by applying gentle pressure and turning it ninety degrees counterclockwise. He pulled it all the way down and then let it rise slowly until the doughnut on the inside settled upon the surface of the fuel. Then it was a simple matter to read the depth of fuel off the calibration marks. The fuel at this location was sixty-four centimeters deep.

The maintenance men drove their truck around to the other wing and repeated the procedure, with similar results. The number 2 drip stick underneath the right wing showed sixty-two centimeters of fuel.

First Officer Maurice Quintal entered the cockpit of Aircraft 604 to find Ouellet and Bourbeau studying the blue-covered drip manual, the book that converted their drip stick readings into meaningful numbers. As he stowed his flight bag by his seat, Quintal heard one of the mechanics ask, "What is that number?"

"Can I help?" Quintal offered.

Bourbeau knew that in order to make any sense out of the drip stick measurements, he had to find the page that corresponded to the attitude measurements. He leafed through until he found the page to use when the aircraft's nose was level. "That's the right page," he announced. "Okay, we're reading sixty-four." He ran his finger down a column of numbers until the drip stick measurement of sixty-four centimeters correlated with 0.5 degrees of wing angle. The drip manual indicated that the left wing tank contained 3,924 liters of fuel. The right wing tank contained 3,758 liters of fuel.

Ouellet wrote down the numbers and totaled them, concluding that Aircraft 604, after its flight from Edmonton, had 7,682 liters of fuel remaining in its tanks.

By airplane standards, the 767 cockpit is a spacious place because it was originally designed for three pilots. But when Captain Bob Pearson stepped into the cockpit on this day he had to squeeze his way past Ouellet and Bourbeau, as well as refueler Tony Schmidt, who had also arrived. Normally, Pearson never saw a mechanic in the cockpit. Usually, by the time he arrived, mechanics had already signed the logbook, certifying the aircraft as airworthy.

But today the mechanics were deep in conversation with Quintal, discussing—in a mixture of English and French —the meaning of column after column of numbers in the drip manual.

Pearson stowed his overnight bag in an aft section of the flight deck, placed his briefcase full of flight manuals next to his seat, and then elbowed his way back out to the main cabin for his routine briefing with the in-charge flight attendant, Bob Desjardins.

"What's going on?" Desjardins asked, nodding toward the crowded cockpit.

Pearson explained that the aircraft had come in from Edmonton with a fuel gauge problem. A channel was inoperative, a drip would be performed, and the flight would be delayed for fifteen minutes or so.

Desjardins decided to board the passengers at the usual time to lessen the chance of further delay once the fueling problem had been resolved.

The refueling panel of a 767, located in a compartment beneath the left wing, holds an array of computerized controls that normally do everything but hook up the hoses. The refueler simply dials in the required number of kilograms, pushes a button, and the airplane quenches its own thirst.

But on this day the array of gadgetry on the refueling panel was blank, as were the cockpit fuel gauges, so someone had to tell refueler Tony Schmidt how much kerosene to board in order to satisfy the requirements of the minimum fuel concept.

Canadian law requires an aircraft to carry sufficient fuel to arrive over its destination airport with enough remaining to enable it to circle at low altitude for at least thirty additional minutes in case of unforeseen delay. For a 767, that legal minimum is 2,200 kilograms more than what is required to reach the destination. Air Canada routinely added another small "fudge factor" to guard against unexpected contingencies, such as the need to maneuver around weather.

Under normal conditions, Flight 143 left Montreal with a relative dearth of fuel, for it needed only about 1,700 kilograms of kerosene to reach Ottawa plus the 2,200 kilograms of reserve. Today, however, it required much more. The revised flight plan that would enable it to reach Edmonton without refueling in Ottawa called for a total of 22,300 kilograms of kerosene, including 300 kilograms for taxiing, the 2,200 kilogram reserve, and Air Canada's fudge factor.

Refueler Tony Schmidt did not know where the aircraft was headed, nor was it his concern. His job was to make sure the tanks were filled with the 22,300 kilograms of kerosene called for on his fuel ticket. He could not simply dial this number on the fuel panel, for the gauges were blank. Instead of relying upon the airplane's fuel gauges, he would have to use the gauges on his truck, and this presented a problem. His truck measured liters, not kilograms. He had to pump in a specified *volume* of fuel to reach a desired *weight* of fuel. He had to count apples, so to speak, in order to arrive at a specified number of oranges.

He and the mechanics now faced a problem reminiscent of a schoolchild's mathbook: an airplane's fuel tanks contain 7,682 liters of kerosene, left over from the previous flight. How many *liters* must be added so that the tanks contain a total of 22,300 *kilograms*?

Mechanic Rod Bourbeau attempted some calculations and ran headlong into the confusion that many Canadians had experienced during the past few years of the changeover from the imperial system to metrics. Seeing Bourbeau's frustration, First Officer Quintal offered to help. "Well," he said, "the number of liters times the weight of a liter will give you kilograms, right?"

It was the job of the refuelers to calculate, thrice daily, the ratio between the volume and weight of fuel. Since the weight of kerosene varies due to temperature and other conditions, one must know the current ratio in order to calculate properly. Seeing Schmidt at the rear of the cockpit, Quintal asked, "What is the specific gravity?"

Schmidt replied, "One point seven seven."

That number sounded familiar to Quintal. He had seen it on many fuel slips. If they were dealing in imperial gallons, the number would be about 8.0 pounds per gallon. In the States, with *their* version of the gallon, the number would be about 6.6. But for liters, 1.77 definitely sounded familiar.

As the men struggled with their math, Captain Pearson, having finished his discussion with in-charge flight attendant Bob Desjardins, edged his way back into the crowded cockpit to begin his preflight check. The veteran pilot had to assure himself that it was legal to fly a 767 with blank fuel gauges. In other words, he was savvy enough to indulge in the age-old pilot's custom known as "covering your ass."

He moved past Quintal and the maintenance men, still standing in the middle of the cockpit, and reached across the empty first officer's seat toward the on-board library, which contained numerous manuals required on every flight, including the aircraft operating manual, the Quick Reference Handbook for emergency situations, a fault reporting manual with code numbers for the reporting of mechanical problems, a manual detailing the layout of the circuit breakers (for use by maintenance personnel, but not pilots) and the volume that now commanded his attention, the Minimum Equipment List, or MEL.

There are three ways to ground an Air Canada flight. If weather is unacceptable, either the captain or the Flight Dispatch Office in Toronto can cancel. For mechanical reasons, a flight may be grounded either on the order of the pilot or Maintenance Control, and here the lines of power are fuzzy, setting up a critical and somewhat adversarial relationship. A pilot cannot fly an airplane unless it has been cleared by maintenance personnel. Once maintenance has cleared the airplane for flight, the "go/no go" decision is the captain's. A captain retains ultimate responsibility for his aircraft—no ifs, ands, or buts. Yet there is subtle but real pressure upon a pilot to do whatever is necessary to fly. A jetliner is a huge investment that produces a profit for the airline only when it is in the air.

Many pilots believe that maintenance personnel have the power to certify an airplane as airworthy, even when this judgment is contraindicated by the MEL. Pearson could remember just such an incident years earlier, in Saskatoon, Saskatchewan. A maintenance clearance in opposition to the MEL cannot force the pilot to fly, but an overly cautious pilot who grounds too many flights for what others perceive as trivial reasons is likely to find himself grounded.

"To ground a flight," Pearson was known to remark, "you have to have a damn good reason."

Thus, the MEL was not always the final authority. Rather, it was an evolutionary document, growing as pilots and mechanics gained sophistication concerning a particular type of aircraft. The MEL of the 767 reflected the newness of the craft; some pages were blank except for the notation TBA, "To Be Announced," awaiting the judgment of experience. Some of the procedures outlined in the MEL were actually impossible to perform. For example, one section detailed a step-by-step procedure for determining the fuel load when one fuel gauge was unserviceable. To do so, one was directed to begin calculations by noting the amount of fuel left in the tanks from the previous flight, as recorded by the flight management computer. The problem with those instructions was that once the aircraft reached the gate and the pilots shut down the engines, electrical power was switched to a ground-based source, causing a momentary interruption of current, and erasing the memory of the flight management computer.

All told, some 55 changes had been made in the MEL in the three months that Air Canada had been flying the 767. In fact, there were three versions of the MEL, one for pilots, one for mechanics, and the Boeing Master MEL, used by Maintenance Control. It was, therefore, somewhat difficult to view anything in the onboard MEL as definitive.

At this moment, however, that information was all that was available to Pearson. He focused his attention on the section he was referred to by Yaremko's logbook entry, item 28-41:02, concerning the two channels of the fuel quantity processor. He read:

One may be inoperative provided fuel loading is confirmed by use of a fuel measuring stick or by tender uplift after each refueling and FMC fuel quantity information is available.

Pearson experienced a moment of uncertainty. He did not really understand how the fuel quantity processor worked, nor was he required to know. He had no idea what channel 2 was, or what the consequences might be if it was inoperative. What he did know was that he had three fuel gauges in front of him, one for each wing tank and one for the smaller reserve tank in the belly, and all three were maddeningly blank.

The drip could supersede all that. What could be more accurate than a bloody drip? he thought. Suppose the oil quantity indicator on the dashboard of your car differed from the reading on the oil dipstick. Which reading would you believe?

Despite the fact that his friends had just flown this airplane in from Edmonton in what Pearson could only assume was the same condition and despite the fact that the Flight Dispatch Office had noted the snag on the flight plan and still authorized takeoff, Pearson held in his hand a document that was unequivocal in its message: It was illegal to fly the aircraft in its present condition.

Musing upon this, Pearson continued with his preflight routine. He flipped the three inertial reference navigation switches to the NAV position. These laser-stabilized, gyro-operated components were the heart of the sophisticated onboard navigational system that allowed the 767 pilot to plan his own route through the skies. In the aft section of the cockpit, he checked to see that the necessary emergency equipment was in place.

Then he eased himself into the pilot's seat in the left front of the cockpit and slid it forward to a comfortable position. He focused momentarily upon the blank fuel gauges near the center of the overhead portion of the instrument panel, but relegated that problem to a corner of his mind as he checked out the other systems. He studied his route charts, adjusted his headset, maneuvered the rudder pedals to a comfortable position, made sure his window was closed and locked, and confirmed that his life vest, smoke goggles, and oxygen system were in place.

One of the maintenance men broke away from the others and approached the rear of the pilot's seat and Pearson used this opportunity to voice his doubts. He turned and declared, "We're not legal to operate in this configuration . . ." he pointed to the blank fuel gauges ". . . with all of the fuel quantity indicators unserviceable."

"Yes, Captain, it is legal to operate," the mechanic responded. "The aircraft has been cleared for dispatch by Maintenance Control."

There was a breakdown in the redundancy system. In the old three-pilot cockpit, the flight engineer bore the responsibility of certifying the fuel load. In the new two-pilot cockpit, the pilots had been instructed that fuel calculation responsibility now devolved to Maintenance Branch, but the maintenance men had not been trained to perform the calculations. They assumed that, in the absence of a second officer, one of the other two pilots would handle the task. Thus, Quintal came to realize, fuel calculation "winds up being the job of an empty seat."

On this day, the resulting confusion found mechanic Rod Bourbeau remaining busy for many minutes, laboring at multiplying 7,682 liters by 1.77, the figure that sounded so familiar to Quintal. Indeed, it sounded familiar because it was the multiplier used to convert liters to imperial pounds. Its use was routine on all types of Air Canada airliners—except the metric-calibrated 767. The proper multiplier to convert liters to kilograms was 0.8, but neither Quintal nor Bourbeau was aware of that. Each assumed the calculation to be the responsibility of the other; neither had been trained to do it properly.

Quintal could see that Bourbeau's multiplication skills were rusty. Although he believed the job was not his, he offered assistance. "I'll give you a hand," he said.

Off to one side of the cockpit, Ouellet took a stab at his own calculation, working on a bit of scrap paper drawn from his pocket. He started his figures too close to the left-hand side of the page, so his numbers became jammed together at the edge of the paper and he had difficulty lining up the columns properly. Frustrated, he started over. Of the three men, Ouellet was the only one to realize that this calculation, if he could get it right, would only tell him the number of pounds of fuel. He would then have to make another conversion from pounds to kilograms, and offhand, he did not know the proper conversion factor.

One of the mechanics finally told refueler Tony Schmidt how many liters to board, but the calculation was based upon the wrong multiplier, the familiar 1.77 that converted liters to imperial pounds. And since an imperial pound is only about half the weight of a kilogram, Schmidt was unknowingly instructed to fill the tanks with only half the necessary fuel to reach Edmonton.

Grievous as the mistake was, there were numerous opportunities to rectify it before Flight 143 reached the point

110

of no return some 41,000 feet over Red Lake, Ontario. One of those arose after the refueling process was completed and Ouellet and Bourbeau were required to conduct a second drip to confirm what they now believed to be an adequate load of fuel. Ouellet made his way back to the cockpit to report to the captain.

By now Pearson was growing agitated, partly because of the delay, partly because someone from maintenance had told him the aircraft was cleared to fly with blank fuel gauges in apparent contradiction to the MEL, and partly over the difficulty these mechanics experienced in performing simple arithmetic.

"Show me your figures," he said gruffly. "You're going to have to prove it to me."

Ouellet leaned forward in the space between the two pilots and exhibited a scrap of paper with his calculations scrawled onto it. According to his figures, the right wing tank now held 6,017 liters of fuel; the other contained 6,234 liters. From Schmidt's signed uplift sheet, Pearson saw that the figures had been converted by multiplying by a factor of 1.77.

To Pearson, as to Quintal, 1.77 was a common number, similar to the one inserted on every refueling slip by the refueler under the heading "specific gravity." He did not know that it converted liters to pounds. Rather, like Quintal and Bourbeau, he made the tacit assumption that it converted liters to kilograms, since that was the number the mechanics had used, and since he believed it was their responsibility to calculate the fuel.

The captain pulled a small hand calculator from his flight bag and double-checked the mechanics' math. Multiplying the right tank load of 6,017 liters by the specific gravity of 1.77, he computed the total as 10,650.09. He assumed that was *kilograms*, when it was actually *pounds*.

Similarly, the left wing tank contained 11,050.11 pounds, but the captain assumed they were kilograms. Adding these figures in his head and rounding them off, he determined that the aircraft was loaded with 21,700 kilograms of fuel, and he had no fuel gauges to tell him otherwise. Had he known the proper kilogram conversion factor of 0.8029 he would have calculated the load at 9,800 kilograms, less than half of the fuel required by the flight plan.

Ouellet was not thinking in terms of kilograms, pounds, or anything else. He serviced flights from Montreal to Ottawa all the time. Unaware that the pilots were planning to fly to Edmonton without refueling in Ottawa, he was befuddled by the excessive concern. He knew that the aircraft had far more than enough fuel to reach Ottawa.

Pearson checked the flight plan, which called for 22,300 kilograms of fuel. They were short. What's more, he was concerned that the load was improperly balanced.

"Well, your calculations look mathematically correct," Pearson said to Ouellet.

"Well, now you can go," Ouellet declared.

"Just a minute," Pearson replied. "It seems to me that there is an imbalance between the two tanks. There is an imbalance of 400 kilograms."

Under normal takeoff conditions, it is permissible for the fuel in one wing tank to outweigh the other by as much as 1,100 kilograms. In flight, pilots could correct this easily, cross-feeding fuel between the wing tanks and monitoring the balance via the fuel gauges. On this flight, obviously, they could not monitor the fuel balance.

"We cannot go because we haven't got individual gauges," Pearson said. "Therefore we cannot balance the fuel in flight. Besides which," he added in a forceful tone, "we look to be below our minimum still. We sure as heck are not going to leave with less than our minimum so you will have to get the fuel truck back."

Flight attendants Annie Swift and Danielle Riendeau wandered into the cockpit, killing time during the delay. Swift glanced out the window, hoping for a glimpse of her boyfriend in the 727 that was pulling up at the adjacent gate.

The flight attendants realized that the aircraft was experiencing some sort of fuel problem and Riendeau quipped, "Wouldn't it be funny if we ran out of gas?" She chuckled at her own joke before realizing that it fell flat. She thought, I have to get out of here because there are too many people in here.

Swift, too, realized their presence was intrusive. She did not know Maurice Quintal, the first officer who was replacing Paul Jennings on this flight, but she had flown with Captain Pearson long enough to appreciate his normally lighthearted mood. This is very strange, she thought. I feel a definite tension in here. "Come on, Danielle," she said to her friend. "I think we're in the way."

The flight attendants left the cockpit to return to their stations.

Ouellet supervised the loading of the additional fuel and then conducted the drip procedure for a third and final time. He now calculated that the aircraft was loaded with 12,525 liters of fuel.

Bourbeau wanted to double-check. "Is it the normal amount of fuel that you put on that flight?" he asked Schmidt.

"It's much more," the refueler replied.

Bourbeau was satisfied. Like Ouellet, neither he nor Schmidt knew that Air Canada had requested enough fuel to reach Edmonton, and had no plans to take on additional fuel in Ottawa.

Returning to his truck, Bourbeau received a radio call from the Maintenance Coordinator in the tower. He wanted to know if Bourbeau was ready to issue a maintenance clearance for Aircraft 604. "Hold on a minute, we're just about finished," Bourbeau said. At that moment he saw Ouellet leave the fuel truck and head toward the cockpit. "Are you just about finished?" he called out.

"Oh, yes," Ouellet replied. "You could give the clearance now. I'm just going in the cockpit to tell the captains the last figure."

"We're just about finished," Bourbeau reported over his radio. "In a second we'll give you the final clearance."

Ouellet then encountered men waiting with the tractor to push Aircraft 604 away from the gate. "How long is it going to take?" one of them asked impatiently.

Ouellet did not reply. He was working as fast as he could.

Pearson and Quintal had nearly completed their takeoff preparations when Ouellet returned to the cockpit and announced, "The fuel is on board and the tanks are balanced. We are set."

Pearson noted on page one of his flight plan that the aircraft carried 22,600 kilograms of fuel, 300 kilograms above its required minimum for both legs of the trip to Edmonton.

But neither pilot was entirely satisfied; there had been too much confusion on the part of the mechanics. Nevertheless, the figures had been checked and rechecked. The tanks had been dripped three times. The aircraft had been allowed to fly from Edmonton to Montreal with blank fuel gauges—or so Pearson and Quintal both thought. One of the mechanics had told them it had been cleared by Main-

tenance Control. They had received maintenance clearance via radio, a requisite when dispatching from either
of Air Canada's main bases, Montreal and Toronto. And
Aircraft 604 had been certified as airworthy by a mechanic's signature in the logbook. If they balked now they would
have to explain their reluctance to an airline that lost a
great deal of money when an airplane was grounded.

Quintal was uncomfortable.

Pearson thought, I have been as prudent as possible.
What more can I do, eh?

In-charge flight attendant Bob Desjardins spoke briefly
with one of the mechanics as he prepared to close the cabin
door. A pilot himself, Desjardins was concerned about the
refueling complications that had delayed departure. "We
better have more than enough," he said.

"You've got more than enough. You can go all the way
to Vancouver," the mechanic replied, referring to Canada's westernmost city, some 800 miles beyond Edmonton.
Then he berated Pearson for his over-concern with the
fuel situation.

Pearson and Quintal found their misgivings alleviated
somewhat as they finished the careful routine of preflight
checks and discovered that the marvelous technology of
the 767 had come to their rescue. The heart of the instrument panel is the flight management computer, with twin
consoles and CRT monitors, one for the captain and one
for the first officer. It was here that Quintal now directed
his attention.

Normally, the flight management computer would receive moment-by-moment status reports from the fuel
quantity processor and keep its own running total of fuel
consumption, backing up the gauges. But today Quintal

had to program the flight management computer manually, punching in the amount of fuel on the plane, as reported to him by the mechanics. The first officer noted the meaning of the prompt boxes that illuminated the screen of the flight management computer. The Boeing 767 instruction manual dictated: "If boxes appear, a fault exists in the fuel quantity processor and the fuel quantity must be entered manually."

Here was another conflicting bit of information. Regardless of what the MEL said, Boeing obviously had built the 767 with the ability to fly with blank fuel gauges. The flight management computer had no means of measuring the actual fuel load in the tanks, but if it was told how much fuel was on board at the start of the flight, it could monitor the amount of fuel being consumed by the engines and keep a running total. This redundancy was comforting. They would, indeed, have backup fuel gauges, displayed on their CRT monitors.

Pearson and Quintal received the necessary clearances via radio: from Air Traffic Control, certifying their planned route; from the Load Agency, certifying that all passengers, baggage, and cargo were loaded within acceptable center-of-gravity limits; and from Station Operations Control, which proclaimed: "Flight 143, you have Maintenance Clearance." They received a last minute weather report from the tower. Temperature was twenty-seven degrees Celsius, or eighty degrees Fahrenheit. Winds were light. Visibility was, in Pearson's jargon, CAFB—"Clear as a fucking bell."

Like many captains, Pearson enjoyed sharing the miracle of flight with his first officer, vying in friendly competition—sometimes augmented by a small wager—to see who could execute the smoothest landing. A flip of the coin determined that Pearson would fly the first leg to Ottawa, Quintal the second to Edmonton.

At 5:48 P.M. Captain Pearson released the brakes of Aircraft 604 and a tractor pushed it away from the gate. The aircraft computer system automatically signaled this event, and its precise time, to Air Canada personnel in the airport. By means of a datalink system that ran through Chicago, the information was also sent to Flight Dispatch in Toronto. These were yet other functions formerly controlled by the flight engineer.

Pearson started the engines. Then he and Quintal conducted the "After Start" checklist, assuring themselves that the engines were performing properly. As Pearson taxied the aircraft to runway 28 the two pilots completed the "Before Takeoff" checklist and received immediate clearance from the tower.

Flight 143 was cleared for takeoff at 5:54 P.M. Pearson opened the engines to full throttle, heard from behind the roar of awesome power, felt the mighty aircraft begin its roll and surge forward, as though it had a life of its own —and knew its natural environment. Racing down runway 28, the aircraft was sucked into the sky by powerful, unseen forces.

Streaking upward from a takeoff speed of 120 knots, increasing to 300 knots within minutes, climbing 6,000 feet per minute at a sharp twenty-five-degree angle, Pearson left the tribulations of the past hour on the ground below.

As Flight 143 cruised easily through the skies of this late Saturday afternoon on the brief—nineteen minutes long —flight to Ottawa, an amber warning light in the cockpit illuminated suddenly. It provided yet another concern for the pilots but was unrelated to the previous fueling questions. It was the bleed valve warning light for the right engine, indicating a malfunction in the system that powered the pneumatic services. Failure of the system could disrupt cabin pressure and temperature, and reduce the effectiveness of airframe deicing. Since it involved only one of the two engines, it was probably not critical, but the pilots wondered whether they should activate the Auxiliary Power Unit to provide backup pneumatic control.

Pearson continued to hand-fly the aircraft—automatic pilot seemed unnecessary on such a short hop—as Quintal checked the Quick Reference Handbook for the proper response. As long as the warning light illuminated for only one engine, Quintal determined, there was no action required.

After a few minutes the warning light went off, but it did not end the pilots' attention to the problem. Quintal noted the incident in the logbook, and then picked up the microphone to alert maintenance personnel in Ottawa. It would have to be checked before the flight continued to Edmonton.

Meanwhile, Pearson mused about the fuel situation. The failure of the fuel quantity processor played upon his mind, calling up the sagacious side of his personality that is such a necessary facet of the professional pilot's character. He knew that the flight to Ottawa would burn minimal fuel, and he calculated in his head that about 20,000 kilograms would remain in the tanks once they landed in Ottawa—far more than necessary to reach Edmonton safely. However, the flight management computer was powered by electricity generated in flight by the engines. When they shut down in Ottawa the flight management computer would go blank. Before leaving Ottawa they would have to reprogram it, and Pearson thought it prudent to know his exact fuel load before leaving Ottawa so that they could feed accurate data into the computer, rather than relying upon a "guesstimate." Accordingly, as Quintal called Ottawa maintenance to report the illumination of the bleed valve warning light, Pearson said, "We are going to have to reprogram the computer in Ottawa. Tell maintenance we want a complete aircraft drip before we leave."

In Ottawa, after completing the brief procedure for shutting down the engines, Pearson rose to stretch his legs, but he soon found the cockpit invaded with people. Rick Dion, the off-duty Air Canada mechanic on holiday with his family, strolled up to say hello. He spoke with Pearson for only a moment before Air Canada mechanic Robert Eklund stepped onto the flight deck and asked if there were any snags. Pearson sat down and turned his attention to the problems at hand.

As far as Pearson was concerned, the fuel quantity question had been answered in Montreal. The additional drip in Ottawa merely provided an updated figure to be programmed into the flight management computer, as required by the Boeing 767 Operations Manual. He was more concerned with the bleed valve problem, and it was to this point that he would direct his attention during the forty-three-minute stopover.

Pearson detailed for Eklund the experience with the bleed valve warning light. The two men discussed the situation for several minutes, including Quintal and Dion in the conversation.

Meanwhile another mechanic, Rick Simpson, arrived in the cockpit to report the results of his fuel drip. Seeing that everyone else was busy, he decided, by himself, to convert his centimeter readings into volume.

"Is there a fuel log on board?" he asked.

"You mean fuel charts?" Quintal replied.

"Yes, that's what I mean."

Dion was standing next to the on-board library, so he pulled out the drip manual and handed it to Simpson.

As the others continued to discuss the bleed valve problem, Simpson found the appropriate charts in the drip manual and converted the centimeters into liters.

The pilots and mechanics concluded that the malfunction of the bleed valve warning light was an intermittent problem. It had illuminated during the short flight from Montreal to Ottawa, but had gone off by itself. Eklund suggested that they prepare the aircraft for departure and monitor the light as the engines were started. If the light came on, they would then decide upon a further course of action. Eklund would stand on the tarmac near the aircraft where he could see into the cockpit. If Pearson gave him a "thumbs up" sign, he would know that the warning light had remained off.

Simpson broke the thread of this conversation when he reported to Pearson, "You have, reading the figures, 5,681 liters in the left, 5,749 liters in the right tank."

That was not what he wanted, Pearson replied.

"You have 11,430 liters in total," Simpson persisted.

"Those figures can't be right," Pearson said. "So we haven't got enough fuel on board, eh?"

Simpson checked his calculations and showed them to Pearson, but Pearson wanted the numbers in kilograms, not liters. He turned and leaned over in his seat to ask the refueler, who was standing at the door with one foot inside the rear of the crowded cockpit, "What's the specific gravity?"

"One point seven eight," the man replied.

Dion was mildly surprised at this answer. He had not had an occasion to use a fuel conversion factor, referred to by the misnomer "specific gravity," since his time in the Royal Canadian Air Force in the early sixties. Back then, he recalled, the number was approximately 0.78. He reasoned that the new number may have been created by the metric changeover, and it was not his business to raise the

120

issue here, as a passenger in a very busy cockpit. He said nothing.

Pearson took out his circular slide rule and multiplied 11,430 liters by 1.78. "We have 20,000—in excess of 20,000," he concluded, confirming the mechanic's calculations. Pearson, Quintal and Simpson all thought they were talking about kilograms when, in fact, they were talking about pounds. Dion assumed that they were talking about pounds, since he was not qualified to service the 767 and he did not realize that its fuel system was calibrated in metrics. Eklund did not even think about it, for he was concerned with the bleed valve problem. Pearson asked Quintal, "What are we looking for?"

Quintal checked the flight plan and reported that they needed about 19,600 as a minimum for a safe trip to Edmonton.

"Well, we have an excess," Pearson said to Simpson. "We won't require any fuel."

According to the erroneous figures, Flight 143 was now loaded with 20,400 kilograms of fuel, and Pearson programmed the flight management computer accordingly. Quintal wrote the number on page two of the flight plan. All available information indicated that they carried 800 kilograms of fuel more than the safe minimum.

Before he left the cockpit, Simpson read in the logbook that the circuit breaker for the number 2 channel of the fuel quantity indicator had been pulled and tagged. Looking above him, he immediately spotted the breaker, covered by yellow maintenance tape marked "inop." It was not pulled, however, as indicated in the logbook, and this gave Simpson an idea. Sometimes an intermittent snag can be repaired by simply recycling the circuit breaker. He reached up, disengaged the circuit breaker, and pushed it quickly back into place. Then he reached over and pressed the fuel test switch, hoping that the digital readout panels of the fuel gauges would register a series of "8s," indicating that they had regained their functional ability. But nothing happened, so Simpson—unaware that Conrad Yaremko in Edmonton had brought the fuel gauges to life by pulling off the circuit breaker, or that Jean Ouellet in Montreal had killed the fuel gauges by pushing it back on—left the circuit breaker as he had found it, in the on position.

At 6:58 P.M. a tractor pushed Flight 143, now only eight minutes behind schedule, back from the gate. Quintal, relishing his chance to fly this long leg to Edmonton, assisted in starting the powerful engines, as both pilots paid careful attention to the bleed valve warning light. It did not illuminate.

Standing on the tarmac, Eklund saw Pearson signal thumbs up. The bleed valve warning light had remained off. Flight 143 was ready to go. Before Dion left the cockpit to join his family in the passenger cabin, Pearson invited him to return after dinner.

Normally a 767 does not take off at full throttle, as the engines generate power to spare. On this day, however, Quintal decided to enjoy a full throttle takeoff. At 7:05 P.M. he applied full thrust to the engines and the aircraft inched forward, then picked up speed exponentially. It hurtled west, toward Edmonton, and fairly leaped into the sky under the direction of Quintal's hand. It rose at a breathtaking rate of more than a mile a minute. Neither pilot noticed anything unusual. With a load of only sixty-one passengers, both Pearson and Quintal expected the airplane to feel light, and this sensation was compounded by the full thrust takeoff. There was, therefore, no cause to believe that Flight 143 might be even lighter than predicted—lighter by some 10,000 kilograms of Type A-1 kerosene jet fuel.

The flight plan called for a cruising altitude of 39,000 feet. Due to the light load, Pearson knew they could cruise at 41,000 feet, saving fuel. As they climbed out of the traffic pattern surrounding the Ottawa airport, Pearson called Departure Control to request clearance to the higher altitude, and it was granted quickly. Pearson also asked permission to fly a direct line to Edmonton, rather than to follow the slightly circuitous paths of the commercial airways. Since the 767 could cruise well above other jetliners, this, too, was granted.

For safety's sake, the flight plan for this longer leg included numerous checkpoints, designated as Timmins, Armstrong, and Red Lake—all in Ontario—and Saskatoon, Saskatchewan. As Quintal guided the plane over each of these points, Pearson would check the fuel load as calculated by the flight management computer and compare it to the minimum fuel load listed on the flight plan.

The first checkpoint, Timmins, appeared far below at 2354 GMT, or 6:54 P.M. local time. On the flight plan, Pearson noted routinely that Flight 143 was cruising at an airspeed of 470 knots at 41,000 feet, encountering a headwind that reduced actual groundspeed to 396 knots. The CRT of the flight management computer indicated that there were 16,000 kilograms of fuel remaining on board. Pearson wrote that number on the flight plan, next to the calculation that proclaimed the fuel minimum over Timmins to be 13,100 kilograms. They were well within the limits.

Later, as Flight 143 soared over Armstrong, Ontario, the second checkpoint, seven minutes early, Pearson noted on the flight plan an airspeed of 469 knots and a groundspeed of 410 knots. Then he checked the fuel figures, placidly unaware of the drama that lay ahead, or the tragedy of errors that led to it.

In retrospect, there would be so many "ifs":

. . . if the single tiny joint in the fuel quantity processor had been properly soldered . . . if a replacement fuel quantity processor had been available . . . if, in Edmonton the night before, mechanic Conrad Yaremko had penned a less ambiguous entry into the logbook . . . if, in Montreal, mechanic Jean Ouellet had not activated the circuit breaker for channel 2 of the fuel quantity processor . . . if one pilot or one mechanic had been charged with and trained for the responsibility of calculating the fuel load . . . if any one of the numerous individuals involved in the refueling process had realized that they had converted liters into pounds, rather than kilograms . . . if, in Ottawa, mechanic Rick Simpson had left the channel 2 circuit breaker in the off position . . .

. . . then Pearson, flying over the Armstrong checkpoint, would have read a different message in his numbers.

But here over Armstrong, where the flight plan called for a minimum fuel load of 10,200 kilograms, the flight management computer reported a comfortable excess of 13,400 kilograms.

We've got bags of fuel, Pearson thought.

14

Flight 155

"How could this happen?" First Officer Gilles Sergerie's question hung in the air, unanswerable.

In the cockpit of Flight 155, following thirty minutes behind Flight 143, the air grew increasingly stuffy. Sergerie snuffed out his cigarette, but lit another one immediately. Normally, he was a light smoker.

It would be years before Sergerie's question could be answered, before investigators could unravel the tangled yarn of circumstance. At the moment there was a far more important question and Sergerie now voiced it.

"What can we do? How can we help?" he asked his captain. There was no reply.

15

The Cabin

In-charge flight attendant Bob Desjardins, briefing his colleagues, had barely gotten the words, "Don't worry," out of his mouth when the no-nonsense voice of Captain Pearson came over the PA system: "Would the in-charge come to the flight deck immediately!"

Now all six of the flight attendants stiffened and exchanged wide-eyed glances. They were alarmed by both the content of the message and the urgency of its delivery.

A red flag flashed in Swift's mind, spelling out in capital letters: BIG TROUBLE!!! She remembered the tension in the cockpit prior to takeoff, and the questions about the fuel supply. We're running out of fuel, she realized. What else could it be?

"I'll be right back," Desjardins snapped. "Gather your emergency cards and go over them."

As he walked quickly forward through the left aisle, toward the flight deck, his mind spun. What can this be? he wondered. He just told me we were diverting to Winnipeg. He just told me we have everything under control. Why does he want me back? What the hell is going on?

* * *

The passengers had been informed that they were diverting to Winnipeg. That announcement had been delivered in a matter-of-fact tone, consistent with airline routine. But this new transmission over the cabin speakers—Captain Pearson's brusque summons of Desjardins—sounded the first note of true alarm.

The words were still ringing in Pat Mohr's ears when she saw the dark-haired, uniformed young man sprint forward past her seat near the galley.

"I'll bet somebody threw up," Pat's daughter Heather said, glancing up from her comic book.

"Maybe so, maybe somebody is sick. . . ." Pat agreed, but every instinct in her screamed that this was not the case. She realized that she was very frightened, and she struggled to mask her anxiety. If I remain calm, Heather will, too, she said to herself. But if I panic . . .

Two aisles across to the right, Ken Mohr's mind was also racing. What in the world is going on here? he wondered. Fuel problem? Fuel feed problem? Well, on a ship like this there must be two, three, maybe more backup systems. But all the logic in the world could not erase from his mind the image of the purser racing to respond to the emergency call. God! I wonder what is going on, he thought.

The eyes of Pat and Ken Mohr met and locked in bewilderment and concern.

Twenty-eight-year-old Pauline Elaschuk had been puzzled by the earlier announcement, but not alarmed. There were problems with the "fuel flow control monitors," or something like that, so they would go to Winnipeg to get it fixed. If the pilots are having trouble with the plane, let them worry about it, she decided. I'm busy with my family.

At her side, baby Matthew slept peacefully. She had

nursed him after dinner and then struggled to calm a bout of colic. Now he lay still, on his stomach, on the seat next to the window. Pauline kept a reassuring hand on his tiny back.

Directly behind her, Richard Elaschuk grew increasingly apprehensive, his mind replaying his two recent experiences with emergency landings. Two-year-old Stephen, seated next to him, fidgeted with his seat belt, straining against it in an attempt to look out the window.

Richard felt the aircraft turn south, diverting to Winnipeg, and knew they had begun a rapid descent. He was disturbed by the lack of detailed information, knowing that the pilots would never explain the extent of the problem to the passengers. The 767 was such a quiet-running aircraft under normal conditions that few passengers were aware that the engines had completely stopped. Left in this vacuum of ignorance, all on board could only speculate.

Turning to his right, his eyes met those of Bryce Bell, seated in the center rank of seats. Bell shrugged his huge shoulders and attempted a faint smile, but the ghostly white pallor of his complexion belied the attempt at reassurance.

Do I look that scared? Richard wondered. If Pauline turns around, what will she see in me? If I look like him. . . . oh, God . . .

Captain Pearson's message to Bob Desjardins was quick, to the point, and almost unbelievable. "We have a problem. Brief the passengers for an emergency landing. We presume we have a fuel starvation problem. Brief the passengers and the crew for a full emergency landing and get back to me when it is over."

Desjardins tried to digest this incredible message. Run out of fuel? The in-charge flight attendant had to draw deeply upon his years of experience, both as a flight attendant and a pilot. He forced from his mind the unanswerable questions of how and why. There was no time. He had work to do, passengers to protect.

He retraced his route to the rear of the aircraft, vaguely noting in passing the questioning glances of passengers. Some started to speak, but held their tongues. Fear. Palpable fear. That was what pervaded the passenger cabin now, he realized.

Arriving back at the galley, Desjardins snapped, "Get your pink cards. We're going to do a demonstration."

Annie Swift felt adrenaline cascade through her. Her hands shook violently. She could not untangle the string that held her pink card, with its emergency instructions, in its plastic encasement near the left rear emergency exit. One of the other flight attendants saw her difficulty, grabbed the card, and ripped it off the wall.

As the attendants moved toward their assigned stations, Susan Jewett's thoughts centered upon the image of baby Victoria, but she forced them back to the present. She stared at the instructions on the pink card. Get the card, read the card, stow this stuff, get rid of it. . . . where? anywhere? Pillows, I must pass out pillows. . . . an able-bodied passenger to help me with the emergency exit.

The team of flight attendants now positioned themselves throughout the aircraft, Desjardins and Clare Morency at the front of the first-class section, Danielle Riendeau and Nicole Villeneuve at the head of the middle cabin, Annie Swift and Susan Jewett at the front of the elongated final cabin that accommodated most of the passengers. Jewett reminded herself to make eye contact with each passenger, as she had been trained.

Desjardins grasped the PA microphone and began to speak, but was taken aback by the quaver in his own voice.

130

He paused briefly, took a deep breath, and began again. As he spoke, the other flight attendants demonstrated:

"I want you to listen very carefully to the following instructions. Remove your shoes, your eyeglasses, false teeth and anything sharp from your pockets. Make sure your seatbelt is tightly fastened as low down around your hips as possible. Cross your arms and hold the top of the seatback in front of you. Rest your head on your arms."

Desjardins continued to speak, repeating the instructions, amplifying. The flight attendants moved through the cabin, visiting with the passengers, putting each through a dry-run of the emergency landing crouch. When it was time, Desjardins said, he would give the signal and everyone was to assume the position.

Swift and Jewett, standing across from one another, kicked off their own shoes. Their eyes met, laden with apprehension. Swift could not bring herself to accept this reality, but as she and Jewett moved through the cabin, assisting, comforting, and consoling, she realized that the countless safety lectures and the repetitive sessions in the flight simulator were working. Her training kicked in. She was doing her job, as if by rote.

Passengers responded with controlled alarm, suffering the onset of a stupefying sense of helplessness. Minutes earlier they had been relaxing after dinner, enjoying a movie, a drink, reading, talking, laughing, fretting over earthbound trivia, planning the days ahead, reliving vacations, anticipating vacations, fantasizing the faces of their loved ones, dreaming of tonight, tomorrow, next week, next month, the years to come. Now their lives were in jeopardy and no one would tell them why. Some anguished at their impotence. Some cried softly. Others asked the unanswerable questions:

"Are we going to be okay . . . ?"

"What is going to happen . . . ?"

Strange, incongruent messages collided in Pat Mohr's ears: "Prepare for an emergency landing. . . . remove shoes . . . eyeglasses . . . sharp objects . . . dentures . . ." That's silly, she thought. If I wore dentures, I wouldn't take them out. If I'm going to die, it will be with my teeth in.

"What's going on, Mom?" Heather wanted to know, now visibly concerned.

"I don't know, honey, there is nothing we can do. . . . Just . . . do what they tell you to do and hold my hand," Pat said, willing her voice to remain calm. "And pray."

Heather took hold of her mother's hand. Quiet, steady tears fell from her eyes.

This is serious stuff, Pat thought. This is not a game. Another part of her brain countered: They will come on and tell us it is not real. It won't turn out to be as serious as they're making it sound, like when the weatherman calls for twenty-five inches of rain and the sun shines.

"Pray," she repeated to Heather.

Before the flight, Pat Mohr would have described herself as an apathetic Catholic. Looking around, she realized there were no atheists on board Flight 143.

Mike Lord's seat belt tightened once more as the airplane took another sudden dip. Rumors flew. Lord thought he heard something about fuel. Passengers glanced at one another nervously. The plane was eerily quiet.

With rapt attention he watched as the crew demonstrated the emergency landing position. Lord did as he was told, kicking off his shoes and stowing them under the seat, stashing a pen out of harm's way, but impulsively slipping a pack of cigarettes and his lighter into the pocket of his shirt. Hanging tenaciously to his glass of beer, he folded his tray, locking it into place in the seatback in front of him. He put his magazine away, cinched his seat belt even tighter, and glanced out the window. It seemed to him that they were descending at intervals, leveling off and then dropping again. He could feel that they were coming down rapidly, twice as fast as normal—or faster.

Winnipeg? They said Winnipeg. Why? Lord wondered. He was attempting to assimilate all this strange information when flight attendant Susan Jewett rushed by and snatched the beer out of his hand.

Nigel Field's British-born-and-bred unflappability retreated in the face of a sinking sensation in the pit of his stomach. He thought to himself, Is this really it? and for some unfathomable reason he found himself muttering a phrase he had often heard in the vernacular of his teenage children. "What a drag," he repeated softly, over and over, "what a drag."

He could feel the rapid and unmistakable descent that, from his own flying experience, confirmed that this was not a normal landing approach. They were going down. But where, and with what result?

Sitting alone on the right side of the aircraft, forward of the wing, Field came to a resolute decision. He would remain as detached as possible and force himself to become an observer to whatever lay ahead, even if it was his own death. He regretted the single glass of wine he had taken with dinner, for it took the slightest edge off his awareness.

Although the flight attendants remained outwardly cool and professional, Bryce Bell sensed terror behind their practiced smiles. He pulled his briefcase out from under his seat, opened it, grabbed his wallet, and shoved it into his right hip pocket. If they were going to crash, if this was to be some horrid replay of the Cincinnati inferno, he wanted his body to be identified.

Why God? Why today? Why me? This can't be happening, he thought.

From his seat in the middle row, he glanced to the right and could see only cloud cover, no ground. He knew there was still time—how many minutes he could not judge—before this drama climaxed. Now he did an incongruous thing. Against all the instructions of the flight attendants, he unbuckled his seat belt and stood up. Crash or no crash, the wine, the Rusty Nails—the fear—made a last trip to the lavatory a necessity. Flight attendants were in a frenzy now, securing objects in the galley. As he elbowed his way past them toward the lavatory, Bell looked into their faces and saw white.

Moments later, as he returned to his seat, Bell suddenly noticed his feet. He was wearing heavy, high-topped leather boots, part of the uniform of the Western Canadian. They were difficult to remove at any time, but doubly so now as he tried to maneuver his six-foot three-inch frame in the confines of an airplane seat. With cold, shaking hands, he finally stripped them off. That done, he practiced the emergency landing position. Leaning forward, he braced himself against the empty seatback in front of him, then swore silently as it collapsed under his weight. He was unaware that all airline seats are built to collapse in this manner in order to provide extra space when necessary.

Bell moved to the adjacent seat and tried again. Once again the seat in front of him could not handle his weight. Muttering epithets against Air Canada, Bell moved a third time, finally finding a seatback that held his weight.

Nearby was a woman traveling with two daughters. He mumbled a reassuring statement about seeing them on the ground. *Liar*, he said to himself.

Flight attendant Annie Swift found a strength, stowed in some detached compartment of her mind, that enabled her to function despite the presence of dark, chilling fear. The minutes ahead promised unknown terror. How long could Captain Pearson maintain control of the airliner? Would they plummet so rapidly they would experience the agony of sudden decompression? Crash? An explosion? Fire? Disintegration? Despite the visions in her head, she knew that people were depending upon her. Supplies had to be stored. There must be nothing that could turn into a projectile and cause possible injury in the event of violent impact. One of the flight attendants, meticulous by nature, was stowing equipment in the galley, carefully placing each item into its proper container.

"To hell with that," Swift said. "Just get rid of the stuff."

Swift grabbed coffee pots, teapots, bar units, liquor miniatures—anything she could get her hands on—and tossed them unceremoniously into a large trash bin, her mind racing: Oh, my God! What is going to happen? Are we really going to die? She heard a soft but unmistakable *ding*! and glanced up to see that the emergency lighting system had been activated. She knew enough about flight operations to deduce a horrifying conclusion: the emergency battery system was activated because the engines were no longer powering the generators—because they had lost both engines!

She said nothing. She did not wish to alarm the others and she worried that if she put her fears into words, she herself might lose control.

Danielle Riendeau felt numbness creeping over her body and realized that she was in physical shock. As she moved through the cabin, demonstrating emergency techniques, the voices of the passengers seemed muffled and strangely distant. She was aware that her breathing had become shallow and rapid. She was cold.

Her most immediate concern was the elderly French woman, the first-time flyer who had asked her to hold her hand on takeoff. Each time Riendeau drew near, the woman jabbered in French. "Am I doing this right?" she wanted to know. "What's going to happen? Are we going to be all right?" The tenor of the woman's voice forecast hysteria.

Riendeau attempted to reassure her, but it was difficult, for with every passing moment her own fear escalated. She spotted another woman in the front row of this cabin, traveling alone, who seemed more composed, so she led her terrified charge forward and seated her beside the calmer lady. "Could you keep an eye on her?" she asked.

"Certainly."

Riendeau walked away, amazed that her own voice had seemed so controlled. Inside, she was anything but calm. She thought: We're dead.

She walked back to the jumpseat on the left side over the wing, suddenly remembering that she had not yet removed her own shoes. She pulled them off and hurled them beneath an empty seat. Then she sat down, fastened her safety harness, and waited for destruction.

Shauna Ohe's dread increased exponentially. In the course of a few minutes the once-spacious cabin of the luxurious 767 seemed to her to have narrowed into a claustrophobic tomb.

Annie Swift stopped at her side to review the emergency landing position. Ohe asked suddenly, "Aren't you afraid?"

Swift's expressive dark eyes grew even larger than normal. "You bet I am!" she said.

* * *

For Pearl Dion the early part of the flight had been terrifying—as usual—but this turn of events was unspeakable.

Despite the fact that she was a veteran of the Royal Canadian Air Force, despite the fact that she was a secretary for Spar Aerospace in Montreal, and despite the fact that she had been married to an Air Canada mechanic for more than twenty years and had flown frequently, Pearl Dion remained a reluctant flyer. Now all of her previously imagined fears were coming true. She wished that her husband Rick would return from the cockpit.

As she sat in the center tier of seats, her arms around three-year-old Chris, she saw his trusting eyes grow large and bright. He looked to his mother for reassurance that she could not supply. Oh, poor little guy, Pearl thought. What have we done to you?

Joanne Howitt, with three-year-old Brodie on her lap, experienced visions exploding planes, burning cabins filled with panicked passengers, the silent, charred hulk of a dead airliner. She could not erase from her mind the image of Cincinnati.

"It's going to be all right," her husband Bob said, as he seemed to hunch his body ever more protectively about baby Katie, blissfully asleep in her harness against his chest. "Don't worry. It's going to be fine." Even as he said the words aloud, he thought: Shit! You're lying through your teeth.

Joanne's heart pumped wildly. Her skin felt cold. Her hands, trying to reassure Brodie with caresses, were clammy and shaking. The passenger cabin, so spacious when they had boarded, seemed to have collapsed around her. She felt so confined.

137

Glancing to her left to peer across the aisle and out the window, Joanne noticed a man watching her. He was Asian and upon his lap was a thick volume that appeared to be a Bible or some other type of scripture. She silently dubbed him "the Preacher." Their eyes met for an instant, and Joanne felt him attempting to communicate reassurance. Then he turned his attention back to his holy book.

Joanne felt the tears coming, but before they could spill over, she glanced at Brodie in her lap and realized that her little boy, confused and disoriented, was taking his cues from her.

You have to be calm, so Brodie will be calm, she commanded herself. She swallowed hard, took a series of deep breaths, and compelled the tears to recede.

"Pearl, we're going down. This is not just an emergency landing. We are crashing," Lillian Fournier said to her sister.

Pearl Dayment's eyes were glued to the window beside her. Everything outside, far down below, was a blur.

"Pearl, in five minutes we're going to be dead!" As she gave voice to this reality, Lillian felt suddenly, unaccountably serene. She wondered if she was in shock. Now that she had said aloud that they were going to die, she could not bring herself to believe it. She recognized the contradiction and pondered it.

"God will take care of us," Pearl whispered quietly.

Lillian stared at her sister, who refused to take her eyes away from the window. She knew Pearl was praying.

Pearl whispered, "God, please bring us safely to the ground. Please take care of us. Please don't let anyone get hurt. Please, God, help this pilot. Help him bring us safely down. Please, God, be with this pilot. Be with all of us."

16

Winnipeg ATC

"One forty-three, we have lost your transponder return right now," air traffic controller Ronald Hewett radioed in desperation. His eyes continued to search a blank radar screen.

Anxious seconds passed before Pearson's voice came over the loudspeaker: "Ah, 143, this is a Mayday situation . . . and, ah, we require a vector on to the closest available runway, ah, we are out of 22,000 feet on the . . . with both engines. ah, failed, due to—looks like fuel starvation—and we are on emergency instruments and we can only give you limited heading information. We are heading 230 now, ah, please give us a vector to the nearest runway."

This was a bittersweet message. The good news was that Flight 143 still existed and remained in radio contact. The bad news was that it had lost both engines.

Hewett forced fearsome conclusions from his mind. He repeated: "We've lost your transponder return. Attempting to pick up your target now."

Fumbling with dials, Hewett switched to his Winnipeg source, the only one of his eight radar sources that still had the old style of primary radar available, equipment not dependent upon an aircraft's transponder. He watched his screen intently. The first sweep illuminated a blip in the area where Flight 143 should be. After a couple of additional sweeps he knew he had regained contact. For forty-eight seconds that seemed like an eternity, Flight 143 had been off the screen.

"We have it now," Hewett radioed. "Just stand by on the 230 heading." Only now did he have a moment to reflect upon Pearson's information. Without engines, most of the instrumentation was down. That included the transponder and that was why he had lost the signal. He realized that Flight 143 had just received its first lucky break. Eighty percent of the air traffic control centers in the world no longer used primary radar, even as a backup. In any of those centers, Flight 143 would remain invisible.

The downside of this was that, with only primary radar, Hewett's information was minimal.

Flight 143 was having difficulty navigating, and that meant that first Hewett, then arrival controller Len Daczko, would have to talk Pearson and Quintal into Winnipeg.

17

The Cockpit

Pearson began to think out loud. "Okay, what's the best speed here for descent?" Nothing in the Air Canada training or flight manuals addressed the issue of attempting to glide a powerless 767 safely back to earth. From his own sailplane experience, Pearson knew there was an optimum forward speed, one that would keep the aircraft aloft for the longest possible time. If the aircraft's attitude was too steep, the aircraft would lose necessary forward speed, stall, and plunge nose first toward the ground. Too shallow an attitude would bring them to earth far short of Winnipeg.

Pearson "guesstimated" that an indicated speed of 220 knots would maximize their range, and hoped this would allow him to avoid one of the major hazards of dead-stick landing. He did not want to make the fatal mistake of coming in short of the runway. He told Quintal his plan: He would come in high over the Winnipeg airport to make sure that they had sufficient altitude, then circle for landing. If they came in too high on the final run, he would put the aircraft into a side-slip, kill altitude quickly, and correct for the high approach.

A side-slip is a common maneuver in a light plane, but unheard of in a massive jetliner. To accomplish it, Pearson would have to twist the control yoke in one direction, setting the ailerons to produce a turn. Simultaneously he would jam his foot onto one pedal, pushing the rudder into the opposite direction. The opposing forces created by the crossed controls would be considerable, stressing the aircraft, creating severe drag that would result in a precipitate drop in both airspeed and altitude.

Air traffic controller Hewett reported to Flight 143, "We show you at sixty-five miles from Winnipeg . . . and approximately forty-five miles from Gimli."

"We might make it," Pearson said.

He had heard of the town of Gimli, the site of an abandoned Royal Canadian Air Force Base, but he had never been to the spot in the vacation community situated on a neck of land between lakes Winnipeg and Manitoba.

"Forty-five miles from Gimli. That's a long runway," Quintal said.

"Okay," Pearson said, considering this alternative. He asked Hewett, "Is there emergency equipment at Gimli?"

"Negative emergency equipment at all. Just one runway available, I believe, and, uh, no control and no information on it."

"Uh, we'd, uh, prefer Winnipeg then," Pearson replied. To himself he added, I wonder how it is all going to turn out?

He thought they could reach Winnipeg, but he was not sure, and his mind played over the alternatives. If they came up short, should he attempt the age-old aviator's trick of landing on a highway? You can't see the telephone poles and other obstructions, he reminded himself. You can't see low stone walls and ditches. They'll tear the hell out of an airplane.

No. No highways, he decided. If he had to, he would ditch in Lake Winnipeg. Macabre thoughts intruded. We're gonna have a floating restaurant, serve mashed seagull to tourists, play tape recordings of people screaming. . . .

The captain snapped himself back to the present. He and Quintal slipped into their shoulder harnesses to give them added security and control for the anticipated bumpy landing.

Pearson then checked the handful of standby instruments, designed to operate even if the power source failed. He was flying the world's most sophisticated airliner with the aid of only a magnetic compass, an artificial horizon indicator, an airspeed indicator, and an altimeter. The standby magnetic compass was of minimal assistance, for he had to lean far to his right to see it. He decided to eyeball his course instead, focusing on features of the cloud formations below, judging intuitively the degrees of heading change.

Dion was impressed with Pearson's gentle touch on the controls. Each adjustment was made slowly, carefully. He doesn't want to rock the boat, Dion realized.

Once more Hewett's voice broke into the somber atmosphere of the cockpit with a radio transmission. "One forty-three, a couple of questions if you have the time," he said.

"Okay, go ahead," Pearson replied.

"Number one, total number of persons on board and number two, fuel on board when landing, if possible."

The controller was repeating his earlier requests. The information was vital to the rescue teams now assembling adjacent to runway 18 in Winnipeg. But there was no way Pearson could tell Hewett how much fuel he would have

143

on board when landing. The fuel quantity indicators had been inoperative for the entire flight. He strongly suspected that he had no fuel on board, but he could not be sure. So he ignored that part of the question, and in the tension of the moment snapped out the first answer that came to mind.

"I have thirty-three people on board, including the crew," he radioed.

"Okay."

There were actually sixty-one passengers and eight crew members aboard.

As Flight 143 descended into air of greater density, indicated speed, as measured by the quantity of air entering the Pitot tube on the left side of the fuselage, naturally increased. Pearson eased the nose up by degrees to maintain his optimum indicated speed of 220 knots.

Gliding in for a normal landing with the engines set at idle and no appreciable wind conditions, they would lose an average of about 2,500 feet of altitude each minute. But with no fuel, they could not, of course, maintain the engines at idle to offset the drag of the enormous engine pods. And without the use of a vertical speed indicator that normally indicated the rate of descent, Pearson could only guess how fast they were descending to earth. No one had tried this before. No one knew how well a 767 could —or could not—glide.

18

Winnipeg ATC

At the Winnipeg Air Traffic Control Center, controllers scurried about looking for "plates," localized landing maps for alternate airports, such as Red Lake, St. Andrews, and Gimli.

Someone found a cheap cardboard ruler to lay across the radar screen to help the controllers monitor distance, and Hewett remembered that when this equipment had been installed, technical experts had decreed that they had no need for sophisticated distance measuring devices.

When Flight 143 was thirty-nine miles from Winnipeg, Hewett prepared to hand over control to Len Daczko, the arrival controller. Normally he would ask the pilots to switch radio frequency, from 118.0, monitored by all planes in his sector, to 119.5, the frequency Daczko was using. "Can you take a frequency change in about three minutes?" he asked.

"Ah, no, we would rather stay with you, ah, on this frequency," Pearson replied. The captain did not want to disrupt radio communications even momentarily. What if

145

they attempted to switch frequencies and were unable to reestablish contact?

Hewett agreed and, hearing this, Daczko grabbed a standby transmitter-receiver from a shelf above his screen and dialed in 118.0. He took over for Hewett and radioed the latest Winnipeg weather report. There were scattered clouds at 4,000 feet. Visibility was fifteen miles. A four-knot wind was blowing from 350 degrees.

Pearson's voice came in over the airways, "Okay, and how far away are we from the field now?"

"Thirty-five—correction—make that thirty-nine miles," Daczko replied, glancing at the cardboard ruler. The new controller, as had Hewett, now asked for necessary information: "One forty-three, when you have a second, can you give us the fuel and the souls on board, please."

"Ah, about the fuel, we still think we have a fuel starvation here . . . souls on board was twenty-five passengers plus six, ah, back-end crew members plus two pilots." Once more Pearson erred in reporting the number of passengers. It was the only noticeable sign of tension displayed by the veteran pilot.

19

The Cockpit

From his observer's seat behind Pearson, mechanic Rick
Dion remained on alert, not daring to intrude upon the
concentration of the pilots, but monitoring the flow of
events, ready to offer assistance. He was concerned about
cabin pressure. With the engines off, cabin pressure could
not be maintained properly, and Dion hoped they would
descend to about 10,000 feet, the altitude where extra oxygen was unnecessary, where cabin pressure could be
maintained above the critical level without the assistance
of the engines. If the passengers suddenly found emergency oxygen masks dropping in front of them, they might
panic. In most aircraft, that would already have happened.
Dion reasoned that this airplane was new enough to have
remained relatively airtight.

He marveled at the ability of Pearson and Quintal to
remain cool in the face of crisis. Soon, in Winnipeg, they
would attempt a dead-stick landing, a rare and difficult
task to accomplish with a 300,000-pound jetliner.

Thirty-five miles from the airport, near the southern

shore of Lake Winnipeg, Flight 143 cleared the edge of a flat cloud layer. Peering below, Pearson spotted what appeared to be an abandoned airfield, with three runways set in a triangular configuration.

"What is, ah, this airfield right in front of us here?" he asked Daczko.

The controller replied, "It is Netley Airport and it has been abandoned for several years and, ah, ah, I'm not quite sure of its condition right now. It has not been used for years."

"Okay," Pearson said. "I believe we are going to make the, ah, airport okay."

Quintal was not convinced of that. On the flight plan in front of him he had created a descent profile, noting altitude and distance from Winnipeg, as reported by the controllers. Studying those figures he concluded that they had lost 5,000 feet of altitude in the past ten miles. Now, they still had thirty-five miles to go and only 9,000 feet of altitude left. Their rate of descent would slow as they encountered greater air density near the ground, but would it slow enough? Quintal stared at the page, trying to make sense of the data. He could not believe how long it was taking him to perform this simple calculation. Come on, Maurice, he said to himself. Come on!

He finally concluded: We will be twelve miles short! He glanced up from his calculations and saw Pearson, his jaw set, handling the controls like a rodeo cowboy hanging onto a bronco. "We're not going to make it," Quintal said to Pearson. "We'll be twelve miles short."

20

Winnipeg ATC

Len Daczko was compiling his own figures. His radar, operating on a sixty-mile radius, featured rings etched onto the screen, calibrated at ten-mile intervals. Setting a cardboard ruler alongside these, he was able to provide Flight 143 with continual estimates of the distance to Winnipeg. The aircraft's transponder, attempting to function off the auxiliary power unit, had sent a few sporadic altitude readings before it went dead. At a distance of thirty-nine miles from Winnipeg, Flight 143 had been at 9,400 feet. On the next radar sweep it was at 9,200 feet. Daczko calculated the rate of descent at between 2,400 feet and 2,500 feet per minute. That was about double the normal descent rate. The aircraft was falling at a vertical speed of about thirty mph. That rate would decrease at lower altitudes, but Daczko knew that Winnipeg was an impossible destination. By his estimate, on its present angle of descent, Flight 143 would plunge into the ground about ten miles short of the runway.

21

Gimli

Kevin Lloyd down-shifted his metallic red Honda Civic as he negotiated a curve on Number 9 Highway. He watched the line of sports cars ahead of him, participating vicariously in the enjoyment shared by his friends. Two days earlier, on Thursday, he and his wife Sybil had driven here, an hour north of their home in Winnipeg, to lay out the courses for this "fun rally," the perfect lighthearted way to end a serious day of racing.

An employee of Revenue Canada Taxation, the national tax agency, Lloyd doubled as president of the Winnipeg Sports Car Club Holding Company, Ltd. More often than not, weekends found him careening his Civic around a two-kilometer road course in serious competition.

The competitors were predominately men. To compensate, the club often ran fun rallies such as this to include the wives and children. For this day, the Lloyds had designed a vehicular scavenger hunt. Participants had to count the number of fenceposts along a certain stretch of backcountry road, solve a riddle based upon the name of a store on Center Street, and complete similar lighthearted tasks that took them across the countryside.

The route now drew Lloyd north through several hamlets clustered between the lakeshore and a CN Rail line, past the villages of Winnipeg Beach and Sandy Hook, past a private Lutheran summer camp and the forty-five-bed Johnson Memorial Hospital, and into the town of Gimli. He turned right on First Street South and headed toward the lake, crossing five small intersections before arriving at Second Avenue. There, facing him, with its back to Lake Winnipeg, was the fiberglass monument that symbolized the soul of this unique—some would say quaint—community.

It was the fifteen-foot-high Viking Statue, one leg raised, bent, fixed upon a rock, one hand grasping a battle axe. Its head was capped with Viking horns, its stern Nordic gaze fixed upon the town. The Gimli Chamber of Commerce had raised $15,000 back in 1967 to erect the sculpture, and had invited Asgeir Asgeirson, the President of Iceland, to unveil it on July 30 of that year.

In Norse mythology the word *Gimli* means "a heavenly abode" or "a place of peace" or simply "paradise."

It was, in fact, just that sort of haven for a nineteenth-century group of Icelandic settlers forced from their homeland by an accumulation of disasters, including a series of especially harsh winters and a deadly sheep epidemic. The Canadian government, sympathetic with their plight and in search of a hard-working populace to settle and exploit the resources of the vast dominion, authorized the establishment of an Icelandic settlement in the New World. A tract of about 460 square miles in the Interlake region, a peninsula jutting northward between Lake Winnipeg and Lake Manitoba, was set aside for the immigrants.

The original group of 285 settlers landed on October 21, 1875, at Willow Point on the shores of Lake Winnipeg, and in the spring moved three miles north to the present site at

152

Gimli. There, with the approval and financial support of the Canadian government, they founded the Republic of New Iceland. For the next dozen years, the republic remained as a quasi-independent state within the Dominion of Canada.

Local records indicate that the first permanent house in Gimli was built by one Fridjon Fridricksson. It was about that time, in the summer of 1876, that the initial group was joined by a second wave of 1,200 immigrants who now forsook Iceland after volcanic eruptions in the Dyngia Mountains devastated 2,500 square miles of their homeland.

Life was not easy. Palmi Jonsson fell out of a boat and drowned in the Red River. Jon Thorkelsson mistook a toadstool for a mushroom, ate it, and died. Hjalmar Hjalmarsson and Magnus Magnusson disappeared in a snowstorm for three days and lost portions of their extremities to frostbite.

During the winter of 1876–77, settlers battled a smallpox epidemic that claimed 102 lives and subjected Gimli to the hardships of a quarantine, but also produced a legendary romance. In February 1877, Signurdur Kristofersson and Carrie Taylor determined to marry, but there was no minister available and no legal way to import one across the quarantine boundary, located south of Gimli. Undaunted, Kristofersson and Taylor persuaded a Lutheran minister to meet them at the boundary. Standing just inside the quarantine limit as the minister stood outside, the young lovers exchanged vows. The story was retold over the years with so much warmth that many a New Iceland mother named her baby daughter Carrie in tribute.

The immigrants battled starvation in this harsh northern climate, for the farmland was poor and rocky and local food supplies seemed alien. One woman was reported to have burst into tears and screamed that she would "never be able to really love a foreign cow." Everett Parsonage led a hunting expedition to Big Island, but came home empty-handed. Some of the men devised rabbit snares, but the women were reticent to cook the animals because of their resemblance to cats. Furthermore, the Icelanders were not adept at the proper fishing techniques in their new land. Their nets proved either too large or too small to trap the indigenous species. On their first attempts they set the nets too close to shore and caught mainly driftwood. Finally, in December 1877, Magnus Stefansson fashioned the proper equipment and led a fishing expedition out onto Lake Winnipeg—an effort that spawned Gimli's oldest industry and still produces abundant supplies of whitefish, sauger, and the local delicacy, pickerel.

In 1887 the Icelandic settlers voted to become Canadians, joining their newly named Rural Municipality of Gimli with the province of Manitoba. Over the ensuing years the original Icelanders were joined by small waves of other immigrants, principally Ukrainian but also including groups of Germans, Hungarians, and Poles, giving the region a multinational flavor.

Growth was slow. By 1891 the town proper still consisted of only about forty homes. A railway line arrived in 1905 and that helped, opening the area for the first time to summer vacationers. But Gimli did not get public water and sewer services until 1957 and by 1983 the town limits encompassed a mere nine streets, crisscrossed by seven avenues.

In the summertime the population of the area swells from its year-round base of 2,500 to about 10,000, as vacationers from Winnipeg move in, such as the group of sports car enthusiasts who this day scooted their vehicles about. Following the route laid out for their fun rally, they drove down Center Street, running east and west between the Number 9 Highway and Lake Winnipeg, to First Avenue. Here, at what might be construed as the town plaza (a distinction earned largely by the process of elimination), an RCAF T-33 Silver Star jet fighter, designated CAF 239, sat atop a seventeen-foot concrete pedestal between the

154

office of the Eastern Interlake Planning District and the Bake 'n Steak Restaurant. A plaque at its base declared that the aircraft was "presented to the Town of Gimli by the officers and men of Canadian Forces Base Gimli to commemorate many years of friendship and cooperation."

It is a hollowed-out hulk of an aircraft, and a hollow statement, for the departure of the RCAF was a grievous blow to the local economy, just as its arrival had been a badly needed boost.

The base was a product of World War II, opening in September 1943 at a site three miles west of town. There, the RCAF established Number 18 Service Flying Training School and graduated a total of 622 pilots for the war effort. The base provided a cosmopolitan interface for the provincial community, which is to say, numerous Gimli daughters married young RCAF pilots and flew off to see the world, expanding the horizons of the once-closed community.

Most importantly it brought jobs, and those jobs persisted after the war when the base was designated a Reserve Equipment Maintenance Satellite and utilized primarily as a summer facility for Air Cadet training. Jobs increased in the 1950s as Gimli was used to train pilots from Britain, Denmark, France, Italy, the Netherlands, Norway, and Turkey, as well as pilots from both the RCAF and the Royal Canadian Navy. This influx was more welcome to the locals than to the young men and women who found themselves assigned to the former Icelandic outpost. Many took one look at the postage stamp of a town and nodded at the appropriateness of its Air Force nickname, "Grimli."

In 1953 Gimli entered the jet age. Fledgling pilots moved in to learn the new skills required on the T-33 Silver Star jet trainer. Soon the base became known for its aerobatic flying team, the Gimli Smokers.

One of the hazards of living in proximity to an Air Force base is the ever-present possibility of something falling out

155

of the sky. Although Gimli recorded its share of crashes, fatal and otherwise, most of the debris wound up either on the runway or in Lake Winnipeg. Only once did an Air Force jet plummet directly into town. That occurred in July 1970 when a pilot lost control of his T-33 and bailed out. The empty aircraft crashed into the rear of the North American Lumberyard on Center Street, near the Number 9 Highway, coming to rest about 100 feet shy of gas and oil tanks at a Shell bulk station.

By 1967 Gimli was the busiest airport in Canada—commercial or military. The base employed 250 locals and pumped some $4 million annually into the local economy. And there was promise of more. That same year Group Captain M. H. Dooher, base commander, announced plans for a four-year, $10 million building program. The economy was boosted further by the opening of a liquor distillery owned by Calvert of Canada, Ltd., which added another $1.5 million annually to the local payroll.

Only three years later the situation had changed drastically. Citing a general reduction in Canadian military forces, the government announced that Canadian Forces Base Gimli would close down by September 1, 1971.

Daniel Sigmundson, Sr., was the owner of Gimli Construction Co. and mayor at the time, and he recalls the closing of the base as "a helluva battle, a political struggle. It was the only major base the RCAF had ever closed." The mayor headed a committee that traveled to Ottawa to fight for the existence of the base. They lost, but they won a few minimal concessions. The federal government agreed to pay $1.5 million to turn the abandoned base into an industrial park.

It was a tough time for the town. A local historian reported 1971 as "that year during which Canadian Forces Base Gimli gradually folded its wings and stole away, with family after family departing until farewells became so commonplace that the later departures went without the

156

traditional ceremonies and parties, because so many of their good friends had already gone."

By 1983 the economy of Gimli limped along. There was the distillery, of course, and Olson Bros. Dockside Fish Products, and a somewhat invigorated summer tourism industry, thanks to government reparations. With federal money the residents had constructed, adjacent to the jet fighter, the Gimli Museum Complex, featuring the Icelandic Museum Gallery, the Fishing Village Museum, and the Ukrainian Museum. Straining for notoriety, a plaque in the Museum Complex announced that Gimli is the Purple Martin Capital of Manitoba.

Residents of Gimli had devised numerous forms of diversion. Curling was perhaps chief among them, but there was summer theater, cruises, the annual Old-Timers Reunion Ball, a Creative Arts Week, fencing, sport parachuting, and both roller and ice skating.

There were other amenities to life in Gimli, and Kevin Lloyd and his fellow members of the Winnipeg Sports Car Club were enjoying one of the most popular. The Royal Canadian Air Force had spent millions to refurbish the base and resurface the runways shortly before the decision to shut down operations, and that had left a modern facility laying dormant. Barracks were converted to industrial offices. The eastern runway remained open for the use of light aircraft, but the western runway was converted to a racetrack.

Several racing organizations shared the facilities. Some simply used the runway as a dragstrip, but the Winnipeg Sports Car Club had carved out a two-kilometer course in the configuration of a "W," utilizing the former runway as the final straightaway.

Completing their fun rally, the sports car enthusiasts now headed west from Gimli on the three-mile drive to the Gimli Industrial Park, former site of Canadian Forces Base Gimli, where their campers were parked and tents

157

pitched adjacent to the old western runway. It was time to set up the barbecues on the runway, enjoy dinner, relax, savor the extended sunlight of a Canadian summer evening and rest up for the Sunday races.

Colin Nisbet had already spent much of his day in the air, but he planned one more flight to take advantage of the final hour of sunlight on this spectacular July afternoon. He checked out the yellow-and-white Cessna 152 and awaited the arrival of his student. In his casual garb, no one would guess that he was, by education, an M.D. For the past dozen years he had allowed his practice to dwindle, and now he performed only an occasional consultation, freeing his time for the business of flying.

Nisbet and his partner, Danny Sigmundson, Jr., had found opportunity amid local misfortune back in 1971. When the RCAF closed its training base at Gimli, Nisbet and Sigmundson leased office and hangar facilities and opened Interlake Aviation, a flying school that offered an intensified and accelerated training course. Taking advantage of the former military barracks, Interlake could offer boarding facilities for fledgling pilots who desired to acquire skills in a hurry. The twin 6,800-foot runways were far longer than those at most private airports. Gimli offered the further advantage of relatively stable weather. To the south, the Winnipeg airport was close enough for students to practice controlled takeoffs and landings. In the other direction was the Canadian north and its relatively uncrowded skies.

In fact, Nisbet could think of no better location for a flying school, but he sometimes had difficulty communicating that message. On the rare occasions when the Rural Municipality of Gimli received mention in the press, Nisbet railed, "they called it the site of an abandoned Air Force base. It is anything but abandoned."

22

The Cockpit

"How far are we from Gimli?" Pearson suddenly asked controller Daczko.

"Ah, you're, ah, approximately twelve miles from Gimli right now."

"Where is it, on the right?"

"On your right about . . . your four o'clock position at twelve miles."

Pearson made his decision. He still thought he might reach Winnipeg, but he was *sure* he could make Gimli. The prudent thing was to head for a "piece of cement" that he knew he could reach. "Can you give us a vector there?" he asked.

Daczko instructed him to turn right on a heading of 345 degrees.

Pearson labored with the controls. At 0133 GMT, some twenty-four minutes after the first warning buzzer had sounded in the cockpit, the aircraft moved into a sharp right turn, sliding silently under the cloud cover, moving north-northwest along the western shore of Lake Winnipeg, away from the now-dubious security of a major airport toward the unknown of an abandoned air force base on the outskirts of the Rural Municipality of Gimli.

The three men in the cockpit of Flight 143 were interconnected with Gimli, although one was unaware of that fact. Rick Dion's wife, Pearl, had served at Gimli during her tenure in the RCAF, as had First Officer Maurice Quintal.

Captain Bob Pearson knew of Gimli, but he had never noticed it from the air, prior to this day. Nevertheless, a portion of his heritage was there. At the southern end of the twin runways of what was formerly Canadian Forces Base Gimli, situated between the thresholds, sat an old Vickers Viscount, Aircraft 624, rotting away from neglect, one wing removed from the fuselage. In its heyday it had served dutifully in the Air Canada fleet before being sold to Ontario Central Airlines. A physician from Winnipeg finally acquired the title, intending to move it to his retreat on Willow Island and refurbish it as a summer cottage. When his neighbors objected that it was too big for his lot, he abandoned the plan, and the aircraft as well.

As it happened, one of the Air Canada pilots who had formerly flown Aircraft 624 was Robert Owen Pearson.

23

Winnipeg ATC

At the Winnipeg Air Traffic Control Center, administrative supervisor Steve Denike quickly located a Visual Flight Reference (VFR) supplement, which contained the pertinent information concerning the abandoned RCAF facility at Gimli. He handed the plate to controller Len Daczko, then turned to the telephone.

Denike first called the Royal Canadian Mounted Police at the Winnipeg Airport and requested that they contact their counterparts at Gimli. All available police and fire-fighting equipment should report to the airfield. He then called the RCMP detachment at Selkirk, Manitoba, halfway between Winnipeg and Gimli, and told them to send whatever equipment they could spare. He contacted the Accident Investigation Board once more to advise them of the change in destination. He called the Winnipeg control tower to inform them that Flight 143 was no longer on its way to runway 18. He phoned his boss, Emil Bryska, acting center operations manager. And finally, he called the Gimli meteorological information office, hoping to supply Captain Pearson with more specific weather data.

On another line, Denike's colleague Warren Smith called Portage Air Base and Winnipeg Military Base to see if any helicopters could scramble to Gimli with personnel and supplies to aid in the event of a disaster.

On his transmitter-receiver, Gary Reid, the data controller, dialed 122.8, the universal frequency used by VFR aircraft operating near uncontrolled airports, such as Gimli. Reid broadcast an alert, ordering aircraft to remain clear of Gimli.

Ron Zurba, the departure controller, used a hotline to telephone St. Andrews Airport, closer to Gimli, and asked officials there to continue to transmit the alert to all aircraft in the vicinity.

Denike then sent Reid downstairs to advise Air Canada officials of the developing events.

24

Flight 155

Flight 155 had altered its course subtly, so that the passengers would not be alarmed. The captain slowed the airspeed so that he would not overtake the stricken airliner, and he eased into a gentle arc to the right, pushing north toward Gimli. From an altitude of 35,000 feet, First Officer Sergerie peered below, his trained eyes searching for Flight 143, which, to him, would appear as a mere speck He could not see it.

Sergerie crushed his cigarette in the ashtray at the side of his seat and immediately lit another.

He tuned his radio to a frequency band that might be —should be—monitored by any light aircraft in the area, and began to broadcast an alert.

"There is an aircraft in distress," he called. "Please remain clear of the airport."

Cigarette smoke mingled with the ominous words. Sergerie was aware that his heart was beating abnormally fast.

25

The Cabin

Nigel Field, sitting alone in the forward cabin, just ahead of the right wing, could not take his eyes from the window. Everything appeared deceptively normal. Below him he saw Lake Winnipeg—still a long way off, but looming closer by the second.

His concentration was broken by the sound of a voice, beside him and to the left. "You look like a calm sort of fellow, would you mind changing seats? We would like you to sit by the emergency exit over the wing." It was Bob Desjardins. Having issued instructions for the emergency landing, tutored individual passengers and cleared the cabin of obstructions, the flight attendants proceeded to the next duty listed on their pink cards. Each was to find an able-bodied passenger, preferably a young, strong man, and position him next to an emergency exit. Their training on this point was clear. The "able-bodied" was to cover the emergency exit and begin the evacuation of passengers in the event that the flight attendant was disabled—or dead.

"I would be glad to," Field said. He was grateful to have

a task, something—anything—to do, rather than submit passively to whatever lay ahead. He rose and followed Desjardins down to the right side overwing emergency exit.

Once there and strapped in securely, Field watched as Desjardins showed him how to open the emergency window. The purser instructed him in two procedures: If the pilot could find a suitable surface and manage a wheels-down landing, Field, upon a signal from Desjardins, was to throw open the door and deploy the emergency chutes whereby passengers could slide quickly to the ground. On the other hand, if the pilot was forced to ditch the airplane, most likely in the inhospitable waters of Lake Winnipeg, Field was to use another lever that would open the emergency door but not deploy the chutes. Desjardins left Field with the admonition, "Do not open the window until I give you the signal."

Field replayed the instructions over and over in his mind. Still, he could not keep his eyes away from the small, square window, only inches from his face.

It all seemed quite unreal. If we ditch, he reminded himself, we won't need the chutes. His orderly mind struggled to identify a strange, distant, foreign set of feelings that now encompassed him. He felt like an observer in this surreal escapade toward destruction, watching with detachment as his own death crept closer.

"My baby . . . my baby . . ." a young mother pleaded, clutching an infant to her breast.

Flight attendant Susan Jewett's heart ached for the young woman. "We'll be fine," she said. "We're just going to land in Winnipeg, don't worry."

"Are we going to be okay?" an older woman asked, her eyes begging for reassurance.

"Yes, of course we are . . ." Susan replied. She realized

that the passengers were acutely aware of their impotence, their inability to act on their own behalf.

Jewett searched the area for an able-bodied. There was a man to her left, adjacent to the aisle.

"Sir," she said, kneeling beside him, "I need someone to . . ."

". . . No, no," he said, cutting her off, motioning to the child who sat beside him.

Jewett understood completely, and scanned the aisles for another candidate. There was a man, sitting a few rows forward. He seemed strong enough, but he was holding a pair of thick-lensed glasses. She went to him once again deliberately kneeling so that her eyes would be level with his.

"Can you see without your glasses?" she asked.

"Yes," he replied.

"I need someone to help me open the door if anything happens," she said. "Do you think you can handle that?"

He nodded.

Jewett brought him to the rear of the aircraft and seated him on the right. As she gave him a quick rundown of evacuation procedures she noticed her voice growing louder and louder.

Bob Howitt's scientific mind was analyzing every nuance of the scene. His ears, unlike most of the passengers', differentiated between the normal super-silent cruise of the 767 and the total absence of power. He knew the engines had quit, and he attempted to assess the consequences.

As a young teenager, he had been flying in a single-engine Cessna over Ottawa with an older friend. "You can turn the engines off on this thing and it won't fall out of the sky," the man had said. And he had done just that, switching off the power and gliding over Ottawa, explain-

167

ing to the fascinated youngster the principles of aerodynamics.

Howitt knew his physics. He understood that as long as they maintained sufficient forward speed, the configuration of the wings would continue to provide lift. Certainly a jet doesn't have the same glide capability as a small plane, he thought. But we have some sort of stability.

The solitude was broken by the high-pitched whine of the ram air turbine, hanging beneath the fuselage aft of the right main gear. Howitt did not know where the noise was coming from, only that it made him uneasy.

What bothered him most was that he had no control, no way to come to the aid of his wife, his son, or his baby daughter. He turned his attention upon the flight attendants, their faces blanched and their eyes filled with fear. Thinking about the uncertainty of the approaching landing, horrible as the prospect was, Howitt realized that it would provide his only moments of control. In the air he could do nothing. But he must be prepared for any contingency once they landed—hit?—the ground. Think, he commanded himself. *Think!*

Outside, through the window across the aisle to his right, Howitt could see a vast expanse of water that his sense of geography told him must be Lake Winnipeg. Since the lake was on his right, he realized that they were heading away from Winnipeg now, to the north. He glanced down at baby Katie, so helpless and innocent. Going down in the water would be a really bad thing, he thought. How will I take care of Katie?

He attempted to work out the logistics in his mind. He thought that after the emergency exits were opened, the evacuation ramps would deploy. Maybe we can use them as rafts, he reasoned. We have to get out of the body of the aircraft and into open air, but stay on the ramp. The best thing is to wait until a lot of people get out first. They can handle the water. But with the baby, I've got to stay on the ramp.

The adrenaline would not abate As she attended to her duties, Annie Swift was overcome by a sensation shared by many of those around her. A part of her felt as though she were observing herself from some distance. On the outside she was functioning, doing her job. On the inside she had no control over the energy pulsating through her.

Her eyes once again scanning the passengers in her charge, she saw a young girl beckon to her. She hurried over.

"My mother," the girl said, gesturing to the woman at her side. "She usually wears a back brace. Should she put it on now?"

"What's it made of?" Swift asked.

"It has steel rods in it. . . ."

"No, no, don't put it on now."

I must find an able-bodied, she realized. Her eyes scanned the left side of the cabin. There were a few adult men—Bob Howitt, Ken Mohr, Michel Dorais. She did not know their names, but she could see that they were with their families. There was one young man, but he had obviously had too much to drink. I can't use him—God help me if I have to use him. And I don't want to separate a family. There must be someone traveling alone here. Who? Where? She moved over to Susan Jewett's side on the right and spotted a stocky young man sitting alone near the bulkhead that separated the passenger cabins.

Swift moved forward quickly and knelt in the aisle beside him. The young man had tears in his eyes. "We need help," she began.

Mike Lord suddenly realized that she was talking to him.

"You are traveling solo?" she asked.

"Yeah."

"What's your name?"

"Mike Lord."

"My name's Annie," she said. "Mike, I need your help. If anything should happen to me, I would like you to be able to open the emergency door and hold the chute. Do you think you can do that?"

"I hope so. I think so. Sure, eh?"

"Come with me," Swift requested.

Lord followed her to the rear of the cabin and over to the left, where she showed him the location of the aft emergency exits. "Just sit down in this seat." Swift gestured to the aisle seat in the very last row on the left, directly behind Pat and Heather Mohr.

Lord sat, buckled himself in, and asked, "What do you want me to do?"

"When we land, I need about five seconds or so to get the door open and for the chutes to inflate. All I want you to do is block the aisle after we land, and don't let anybody get by you. If anybody should panic or anything, well . . . we don't want a jam-up. Just give me five to seven seconds, then when I say 'Now!' turn around and jump down the chute and stay at the bottom to help the others as they come down. Can you do that?"

"Yeah, no problem."

26

The Cockpit

With the cloud cover now above them and moving off to their right, Pearson and Quintal could confirm visually that they were heading due north along the shoreline of Lake Winnipeg. Pearson aligned the nose of the aircraft with a highway running adjacent to the water. Below was barren farmland composed of coal-black dirt, broken by clusters of vacation cottages and modest houses formed into small hamlets. He knew that one of the towns ahead was Gimli, and that somewhere near Gimli was an airfield, but it was difficult to discern detail in this unfamiliar terrain. A 767 normally does not land at a small-town airport, and both pilots realized that their perspective was askew.

Pointing ahead to a spot about ten miles distant, Quintal introduced Pearson to Gimli.

The captain could see the outlines of the small community. Near the lakeshore he spotted the familiar straight-line pattern of a runway.

"Bob, do you see the airport?" Quintal asked.

"Yeah, straight ahead."

Quintal peered ahead at something that resembled a runway, but having served here in the RCAF, he knew it was not. He pointed to Pearson's eleven o'clock position and said, "Bob, the airport is over there."

"Okay." Pearson had mistaken something else—he knew not what—for the runway. He adjusted the controls gently, maneuvering the plane slightly to the left.

The cockpit loudspeaker came to life with the voice of controller Len Daczko. "Okay," he radioed, "Looks like you are lined up straight for it and I would say you have ten miles to fly."

They were coming in fast and sharp with no flaps or slats to slow their speed and they had no engines to provide maneuverability. Both pilots were acutely aware that what they also lacked was a second chance.

"Okay, fine, ah, we're going to go in there," Pearson reported. "Do you have emergency . . . can you . . . whatever emergency equipment you can get there from the town, will you get it?"

Daczko replied: "Affirmative. Ah, we'll have the . . . ah, call to the local authorities, ah, out there and, ah, ah, we will get everything out we can for you."

On his radar screen Daczko had spotted an unidentified target, possibly a small private airplane, about five miles ahead of the airliner and slightly to the right. Its pilot had either not heard, or ignored, the numerous warnings to clear the area. Daczko reported to Pearson, "I show a target just coming up to your one o'clock position about, ah, ah, I'd say five miles opposite direction. Slow moving, type and altitude unknown."

"Roger."

Pearson and Quintal shielded their eyes from the western sun and glanced forward, finally spotting the other aircraft, a mere dot illuminated on the red glow of the horizon. It was a small private airplane operating visually, not in contact with ATC. Pearson estimated that they would pass over it by at least 3,000 feet.

"We have the field in sight," the captain reported to Daczko. They were close enough now so that Pearson could see the towers of the approach lights, set twenty-five feet apart to signal the approach path. The captain did not realize that there were two runways ahead of him. The light towers were, to him, unmistakable evidence that he had zeroed in on an active runway.

In fact, the old approach light towers had never been removed. They heralded the approach, not to runway 32 Right, the landing site currently used by small aircraft, but to what was formerly runway 32 Left and was now the final straightaway of the two-kilometer racecourse utilized by the Winnipeg Sports Car Club.

27

The Cabin

Shauna Ohe and Michel Dorais reacted differently to the crisis. Dorais grumbled that the instructions being given were inadequate and confusing. His concentration upon the activity inside the aircraft made him unaware of the almost 180-degree course change that had taken them away from their announced emergency destination. We'll land in Winnipeg and we'll get off, he reassured himself. He rejected any other conclusion.

At his side, Ohe felt an odd, peaceful realization envelope her: I love this man with all my heart and soul. What a wonderful way to die, holding this man. I am going to die and I am one of the lucky few who know that I am dying.

Belted securely into the left aisle seat in the back row of the airplane, primed for his job of blocking the aisle until flight attendant Annie Swift could open the emergency exit, Mike Lord was alone with his thoughts:

Wow, my last meal. I had fish for my last meal. How

about that? Fish. Then uglier images surfaced. He knew that few people survived air disasters, but what if *he* survived? I wonder what parts of my body I will lose? he mused. My legs? An arm? If I survive, would I even want to? Will I wind up in a hospital in Winnipeg for a year or so with no legs? Where will they bury me? Or will it be a mass grave somewhere? He thought of his family and friends, of his cousin Debbie—waiting in Edmonton for a plane that would never arrive.

The waters of Lake Winnipeg seemed dangerously close now. He's going to dump it in the lake, Lord said to himself, I'm a dead duck. Mike Lord did not know how to swim.

In the very centerpoint of the cabin, three-year-old Chris Dion stared up at his mother wordlessly, his eyes reflecting both fear and the certain knowledge that his mother would take care of him. Impulsively, Pearl Dion unbuckled his seat belt and pulled him onto her lap.

Susan Jewett, making her rounds one final time, approached with a mild rebuke. "You'll have to put him back in his seat," she ordered gently.

"No, I want to hold him."

Susan regarded the scene in quiet empathy and decided to ignore this point of the safety regulations. What better place was there for a child than in his mother's arms? "Okay," she agreed, and moved on.

Once more Pearl Dion looked at her son. She began to sob quietly. Tears rolled down her cheeks and fell upon Chris's tawny head. Still, the child said nothing.

Boots off, pockets empty, head tucked between crossed arms, the normally loquacious Bryce Bell began a silent dialogue. His son Jonathan's small face loomed before him. He hadn't played with him enough. He had taken him for granted. He hadn't even hugged him or kissed him good-bye this morning. What would this small child, this three-year-old boy, remember of his father?

And Margo? If God would just get him out of this he would be more considerate—treat her better. I'll try to enjoy life more, get out of this damn job and do something worthwhile with the rest of my life. He thought of his parents—they were elderly—it wasn't right to leave them this way.

Bell could feel the plane descending, steadily, unalterably, and quickly, too quickly. It began to oscillate, shaking and bouncing and rattling. He braced himself for the moment of impact, and his silent, lonely prayer continued. Oh my God, please help me, please help me. Get me through this, God, please God, no.

Through the window, into the sunny, clear July evening, he could see nothing but water. Oh God, I hope we go in the water. If we go in the water I won't burn. I don't want to burn. The water will be softer. I don't want to go into the trees. I don't want to burn. Please God, don't let me burn.

28

The Cockpit

Rick Dion discussed the upcoming landing with Pearson and Quintal. Airspeed was a troubling question. Sufficient airspeed was essential in order to maintain lift. If airspeed dropped below the critical point—and no one was quite sure what the point was—the aircraft would stall and plummet to earth. Without engines Pearson's only method of controlling airspeed was the degree of nose-down angle, which he was forced to decrease as they reached the heavier air of lower altitudes. Conversely, too much speed would be a killer, causing the aircraft to disintegrate upon landing or to career off the far end of the runway.

It was this latter problem, too much speed, that most concerned Dion. The minimal hydraulic power supplied by the ram air turbine was not designed to extend either the slats on the leading edge of the wings or the flaps on the trailing edge. On a normal landing these provide greater lift, allowing the aircraft to slow to a speed of about 130 knots. Without them, Pearson would have to touch down at a speed of 180 knots; he might well blow a tire

179

or two, compounding the difficulties. He's going to have his hands full stopping this baby in time, Dion predicted silently.

From his vantage point in the jumpseat behind Pearson, Dion calculated the odds: There is no fuel on board, so that eliminates fire. If he doesn't make it to the runway, there's flat prairie below. If he has to put it down in a field he will land gear-up and the engines will break off, but the central tube of the fuselage will probably keep going intact.

Dion knew that the aircraft was engineered for such a contingency. The engine pylons were built to break away on impact so that the aircraft could slide on its wings. *We have a chance!* He was grateful that Pearl and Chris were seated in the very heart of the airplane, in the middle of the tube, in what he believed to be the safest possible position.

There's nothing else I can do here, he thought to himself. I might as well go back to the cabin. If we don't make a successful landing, well, I'm an able-bodied male. I'll be more help back there. With them.

He looked at Pearson. The captain was perspiring heavily, his posture signaling intense concentration and apprehension. His hands gripped the controls tightly.

"I'm going back, Bob," Dion said.

"Eh."

29

The Cabin

As Rick Dion walked down the left aisle of the passenger cabin his eyes took in a scene of bewilderment and fear. All the passengers had their shoes off. Many were buttressed in their seats with pillows and blankets. Some were already crouched over into the emergency landing position. One woman, still sitting upright, was sobbing.

Heads popped up as he approached. Strange eyes met his and pleaded for answers. Who was this man in the business suit who suddenly emerged from the cockpit? They don't know what's going on, Dion realized. Then he decided, That's probably just as well.

Dion finally reached Pearl, who waited anxiously with Chris on her lap. Before he could sit down next to her she asked, "What's going on?"

Anyone within earshot waited for his reply. He knew he had to be careful with his answer lest he generate panic.

"We're going to land in Gimli instead of Winnipeg," he said, then added, "to get some fuel."

"Have we got enough fuel to get there?" she asked.

"Oh yeah, no problem," Dion lied. But he thought: Please quit talking about it.

Rick Dion now made a conscious decision to keep his shoes on. He knew that the reason passengers were instructed to remove their shoes was to prevent them from ripping the surface of the evacuation chutes. He was an Air Canada employee. He was an able-bodied male. He had every intention of being the last one off this airplane.

Pearl Dayment noticed the *absence* of sound. Here was this huge jetliner careening toward earth, and there were no motors, no jets, no sounds of power. "We'll be fine," she reassured her sister, Lillian Fournier. "God will take care of us."

"I hope you're right, Pearl," Lillian replied. "I hope He hears you. We're going awfully fast. . . ." She glanced past Pearl and out the window. The lake was gone from view now, replaced by grassy fields drawing closer and closer. She had seen enough. She put her head down in her lap and tensed for the inevitable impact. It will be over soon, she thought.

As the attitude of the plane shifted, an overhead compartment shook open and something crashed to the floor, causing Bryce Bell to wince. He took a deep breath and braced himself for fatal impact. When it comes, he imagined, my guts will be with these strangers' guts. We will all be one massive pile of humanity.

They were very low. Through the window to his left he could see golfers, running like hell for the woods. Christ, I can almost see what clubs they are using, this is it. This is it.

He looked around the cabin. Ahead of him an elderly woman, reading glasses perched on the edge of her nose, scribbled frantically upon a scrap of paper; Bell realized she was writing a will. To the other side a young mother clutched an infant to her chest and sobbed quietly. Somewhere a child yelped in pain, complaining of an earache brought on by the rapid descent.

Bell thought: This is a slow way to die.

The few feet that separated Ken and Crystal Mohr from Pat and Heather seemed, to Ken, to stretch. Maybe we should move over closer to them, he thought. Maybe we'd better stay here. Things were happening so fast . . . too fast. He looked at Crystal, who seemed unaware of the drama unfolding around her. She was so young, so innocently curious, that every attempt to get her to keep her head down was fruitless. Ken placed one hand in front of him as a brace and used the other to restrain her head from bobbing up and down. His stomach was in a tight knot. Oh, my God! he thought, I wish we were somewhere else.

Beside them, through the small window next to Crystal, Ken could see that they were dropping rapidly. He took his hand from Crystal's head and pulled down the shade. He did not want to see anymore.

He felt his stomach flip from one side to the other as he repeated to Crystal, "We're going to be all right, sweetie, don't worry, everything's going to be fine. . . . We're going to land in Winnipeg, that's all. . . ." He clenched his teeth and resolved: We will not perish on this airplane, we will not. We may crash, we may be hurt, but I refuse to die and I refuse to lose my family on this airplane. He kept his body rigid, braced for impact. His eyes were open. He waited for the wrenching sound of destruction and the inevitability of the aircraft tearing apart.

His body seemed to hover in a void of space and time.

* * *

Never in her life had Pat Mohr experienced such a rapidly shifting rush of emotions: confusion, concern for her family, cold, gut-wrenching horror over the certain realization that they were about to perish, to die a fiery, brutal death. Then a blanket of sadness engulfed the young mother as she realized that her two daughters would never grow up. With that thought, Pat was astonished to find that she was furious. She did not know at whom this rage was directed, but somebody, somewhere, was responsible for destroying the future of her children.

Eleven-year-old Heather clutched her mother's hand tightly and thought: Is it going to hurt? Quiet, steady tears rolled down her cheeks. She glanced out the window and gasped.

Her mother reached across in front of her and pulled down the shade. She wanted to know what was happening—and she did not want to know. She wanted to see—and she did not want to see.

The cabin suddenly seemed like a quiet, falling submarine. A claustrophobic sense of suffocation swept over her. She did not know when they would hit or what they would hit. She only knew that whatever was going to happen would happen soon.

In the aisle, flight attendant Danielle Riendeau encountered Bob Desjardins, who told her, "We're coming down fast. Go back to your seat and buckle up."

Although she had been flying for a decade, Riendeau possessed only the most elemental knowledge of aeronautical principles. She had no concept that a powerless jetliner might be able to glide. She was convinced that at any moment the plane would spiral to the ground and shatter into a million fiery pieces. We're dead; we are all dead. There is no point. . . . We are all dead anyway, she despaired.

184

Pauline Elaschuk did not succumb to terror until Annie Swift helped her prepare baby Matthew for the emergency landing. Swift pulled two blankets from the overhead compartment. Together the two women formed a cocoon for the sleeping infant. Then Swift showed Pauline how to clutch Matthew to her breast, bend forward, and use one arm to brace herself against the seat in front of her.

"If we hit, you will be thrown forward," Swift warned.

Pauline practiced bracing herself, trying to find the best position for her arms so they would not crush Matthew.

"You're doing fine," Swift reassured, patting her on the back.

As Swift moved on to attend to other passengers, Pauline suddenly felt very cold, void of emotion. She looked at the helpless baby in her arms and kissed him gently.

I might never kiss you again, she said to herself.

Craning her neck to the right, looking behind her, she saw her husband Richard and their two-year-old son Stephen. She whispered—or thought—"I may never see you again," and Richard silently understood. Pauline reached a hand up to the back of her seat and Richard reached forward.

Briefly, their hands touched.

"Susan, you better sit down," one of the other flight attendants said to Jewett, and she complied, taking her post in the very back of the airplane, in the flight attendant's jumpseat on the starboard side of the galley facing forward. Everything is so quiet. . . . It is too quiet, she thought. A sensation of total helplessness descended upon her. I don't want to be an invalid. I'd rather die than be an invalid. God, I wanted to see my little girl grow up . . . how will they manage? Finally she thought: I had ten good years of flying.

135

* * *

Annie Swift returned to her seat at station L2 in the left rear, across from Susan Jewett. The interior of the cabin dimmed as the emergency battery failed. Swift knew that it was designed to operate for a mere fifteen minutes. The light of early evening filtered in through the windows, casting solemn shadows.

"Oh, my God!" Swift said. "What is going to happen? Are we going to get out of this?"

She glanced forward at her friend Danielle Riendeau, seated over the left wing.

Oh, God! Swift thought. She won't have a chance over the wing. If we crash everything will fall on her. God! Why does she have to be there? How will I get her out of here?

Passengers cried softly. The air was thick with a collective, unyielding dread.

30

The Cockpit

"What's the runway like?" Pearson asked.

"It's used, ah, all the time by VFR, ah, aircraft and aircraft of DC-3 type," replied controller Len Daczko.

"Roger. And, ah, th . . . there will be nobody on the runway when we get there? Nothing?"

"I don't know. I can't, ah, tell you for sure."

Normally, Pearson could rely upon the glide slope indicator to tell him whether to increase or decrease power. Or he could program the computers to land the airplane; its three electronic autopilots would act in concert, "voting" on each possible control adjustment to arrive at a consensus that would bring the aircraft down dead center on the runway and apply just the proper amount of brake pressure as preselected by the pilot.

Normally.

For this landing, he—and everyone on board—had to rely upon his accumulated skill and seat-of-the-pants judgment.

Sitting adjacent to his captain, Quintal commanded his

practiced eyes to evaluate the approach, to estimate the distance to the threshold of the runway. He could only guess at the rate of descent. We are going to make it, he decided.

Moments later his mind screamed, We're not going to make it! We're coming in too fast!

31

The Cabin

Nigel Field was mesmerized by the view from the window. They seemed so close to the ground, skimming the trees and bushes, that they must surely strike something at any moment. He imagined how it would feel as the plane cartwheeled and broke apart once the engine made contact with the ground. Field estimated the speed of the shaking, oscillating, wounded craft at well over 200 mph, and his self-imposed detachment gave way to a sense of bitter disappointment mingled with fear. He knew, for the first time, that he and the rest of his fellow passengers would not survive the next few moments.

And this is how it is to be? he thought.

32

Winnipeg ATC

In Winnipeg, in the darkened, closed room of the Air Traffic Control Center, a group of tight-lipped men stood behind Len Daczko, watching his radar screen intently. As one, they ached for the plight of the crew members and the inaccurately reported "souls on board."

Finally the moment arrived, as they knew it would, when Flight 143 dipped below the horizon. Their radar could no longer detect the presence, or existence, of the airplane.

On one sweep of the radar beam it was there as a distinct blip.

On the next sweep it was gone, leaving behind only the remnants of an echo that faded quickly into nothingness.

33

Winnipeg Beach

Robbie and Patti Dola relaxed in lawn chairs in the front yard of their home six miles west of Winnipeg Beach. The young couple had spent a pleasant Saturday afternoon working in their vegetable garden. Now, Robbie had thrown a few hamburgers onto the grill. He sat back contentedly, allowing the aroma to whet his appetite.

Suddenly he heard a strange noise overhead, an eerie, high-pitched whine. Looking up he saw a jetliner, perhaps a half mile off the ground, passing overhead from south to north. The sound was unlike anything he had ever heard emanating from an airliner . . . and it was not coming from the engines.

"There's something wrong with that plane," he said to Patti. "There's no motors."

Patti caught a glimpse of the Canadian Maple Leaf emblazoned on the tail "It looks like an Air Canada plane," she said. She happened to have her camera handy, so she grabbed it, aimed, and snapped a few photos.

The aircraft passed quickly across their line of sight.

193

Robbie jumped from his lawn chair, ran to the back of the yard, and peered around a stand of trees, but the jet had disappeared. It was headed for Gimli, but Robbie was sure it was losing altitude too rapidly. It will never reach the runway, he thought.

Robbie and Patti climbed into their black Ford pickup and drove off, winding along back roads, listening for an explosion, scanning the distance for a fireball and a column of smoke.

34

The Cockpit

Despite the distortion in their perception caused by the approach to a small-town airport, Pearson and Quintal were now convinced that they were too high—far too high. Unless they acted immediately they would overshoot the runway and end up somewhere in the fields beyond, or, worse, in one of the clusters of modest homes north of the airport.

For an instant Pearson considered putting the airplane into a tight 360-degree turn, but feared they might lose too much altitude and fall short. Such a precipitous turn might also disorient him and make it difficult to reestablish visual contact with the threshold.

Lowering the gear would slow them down. "Okay, Maurice, gear down," Pearson said.

Quintal pulled the hydraulic undercarriage selector to the down position. Both pilots waited for the reassuring *whoosh* and *boom*, the sounds of the gear dropping into position, and the controlled vibration caused by the increased drag. But they encountered only silence.

The eyes of the two pilots met, knowingly, and at the same time full of questions. There was not enough hydraulic pressure to extend the landing gear in a normal manner. Could they get the gear down? If they reached the runway, would they have to bring her in on her belly?

Quintal searched in the Quick Reference Handbook for the alternate gear extension checklist. Logically, he turned to the chapter on landing gear, but the index listed no heading for an alternate gear extension procedure. Quickly, he skimmed through the pages of the chapter. "Bob, I can't find it!" he said in desperation. Then he decided to freelance. "I'm going to activate the alternate switch."

"Go ahead!" Pearson replied.

Quintal reached for the alternate gear extension switch, a standby system that pulled out the metal pins that held the gear door closed. The doors, in turn, held the wheels up within the belly of the aircraft during flight. With the pins out, the force of gravity would drop the wheels into place. "Okay, Bob, am I hitting the right switch?" he asked, following his training.

"Yes," Pearson confirmed.

Quintal removed the guard from the switch, then pressed it and they heard a satisfying *wo-o-o-ong* as the wheels fell free, slamming against the doors, forcing them open. They felt the vibration of the increased drag, and they began to lose altitude more quickly.

On the panel in front of Quintal, two green lights indicated that both main landing gear were down and locked. But another glowed amber, warning that the nose gear was partially down but not locked. Quintal immediately knew why. Dropping from their housings instead of being powered into place, the main gear fell sideways. But the nose gear had to push forward, against the wind, and there was not enough time before they reached the ground.

Quintal once again dove for the Quick Reference Handbook, searching for the passage that detailed emergency procedures for locking nose gear into place. By now he knew that, for some inexplicable reason, this information was not in the landing gear chapter.

Hydraulics! he told himself. It's a hydraulic system. Check there. His fingers sped through the book, racing against time. He read: "If you lose your left hydraulic . . . If you lose your right . . ." *No, that's not it.* "If you lose the left and center . . ." *No, that's not it.*

35

The Cabin

Nigel Field's terminal reverie was interrupted by the noise of the undercarriage being lowered. He was astonished that there was still time for the pilot to attempt a wheels-down landing. We must be close enough to some large, flat strip of pavement, he concluded.

In the center of the passenger cabin, mechanic Rick Dion had been listening to the sound of the ram air turbine, hanging underneath the belly aft of the right main landing gear. From the sound of the RAT, he could tell ahead of time whenever the aircraft would make a course correction. When Pearson adjusted a control, the RAT's four-foot propeller responded with a lower pitched hum, straining to provide the necessary hydraulic power to move the ailerons, the rudder, or the elevators.

"Good," Dion said, when he heard the *thud* of the main landing gear as it locked into place. This was an obvious indication that Pearson believed they could make the runway at Gimli, rather than crash-land in a field.

* * *

This is it! Joanne Howitt knew, hugging three-year-old Brodie to her chest, fighting hard to choke back the tears.

"Stay calm and everything is going to be okay," Bob Howitt repeated. "Keep a clear head and we'll come out of this one okay." Joanne looked at her husband and knew that the reassurance was hollow.

Bob tried again. "At least we're coming down over land now. We won't have to deal with water." To himself he said, It could be hard and ugly. The plane could slew and jackknife. Is it physically possible for this plane to catch the nose and flip endwise? We appear to be coming in level and flat, but what happens when we hit soil, or concrete?

Beneath him, he felt the undercarriage drop. He heard it lock into place and felt the aircraft shudder and lose speed due to the increased wind resistance. How much airspeed is he losing? he wondered. Has he made a mistake by dropping those wheels and losing too much speed?

Richard Elaschuk was also attuned to every movement of the control surfaces. Whenever the pilot made an adjustment, Elaschuk heard the grinding *whirr* of motors that moved the control surfaces, and felt the shuddering response of the aircraft.

Richard felt his mind detach from his body. Standing off to one side, he watched as he attempted to comfort two-year-old Stephen, to keep him calm and yet position him for maximum protection. He regarded his own image: You are pale. Your eyes are really wide open. Your mouth is dry. So this is how you feel when you are . . .

36

The Cockpit

Pearson was too busy to notice that the forward gear had not locked. He had another problem on his mind. Intuition and experience told him immediately that the increased drag of the landing gear was insufficient. They would still come in long and, without full anti-skid controls on the wheel brakes, without the ability to apply reverse thrust to the engines, they would careen well past the end of the runway. He had to lose altitude and he had to do it fast.

Pearson turned the yoke to his left, simultaneously jamming his foot against the right rudder pedal. Outside, the ailerons on the trailing edges of the wings responded to the command of the pilot's arms, the left aileron swinging up, the right one dropping down, disturbing the airflow. Normally, this would send the aircraft into a left turn. At the rear of the airplane, however, the rudder swung to the right, held there by the strength of Pearson's foot on the pedal, fighting against the force of the ailerons.

Responding to this cross-control, the aircraft turned sharply onto its left side and lost altitude precipitously, but maintained a forward course toward the runway.

Incredibly, Pearson had the giant 767 in a side-slip, a maneuver unheard of in a jetliner.

37

The Cabin

"Oh!" Pearl Dion gasped, responding to the sudden change to a precarious angle and the accelerated sensation of falling.

Uh-oh, he's losing it, Rick Dion thought. He's slowed to a point where the RAT is no longer effective. The plane is not responding to him. If he doesn't regain control, if the wing tip hits, it will be all over.

The possibility that Pearson was executing a side-slip in a powerless 767 never entered Rick's mind. If he had considered it, he would have concluded that there was insufficient control to attempt it. Crossing the controls puts a tremendous strain on any aircraft. He would not have believed that the RAT could supply enough hydraulic pressure, nor that it was humanly possible for Pearson to muster the strength to hold the controls at opposing angles.

It felt as if the plane was nearly on its side. Out the left-side window *below him*, Rick spotted sand traps on a golf course. "Holy Jesus!" he muttered and his mind gave way to terror and despair. He curled his strong arms around Pearl and Chris.

I wonder what it's like to die, Pearl thought. I'm sure going to find out.

Michel Dorais reacted to the sudden severe sideways angle of the aircraft, looking out across Shauna Ohe and to the ground below, where he saw golfers peering back at him, incredulous expressions upon their faces. For the first time, he faced the possibility of death.

Ohe clutched his hand tenaciously.

"I love you, Michel," she said.

"I love you, Shauna," Dorais replied.

Ohe prayed: "Please, Father, let my children know how much I love them." She thought, This will be a very violent death.

This breathing, this awful breathing, thought Danielle Riendeau, as her chest rose and fell under the bodice of her flight attendant's uniform. I can't get enough air. I am breathing too hard. . . . She waited for the airplane to lurch, nose down. She closed her eyes. They were low, very low, and tilted steeply to the left. I don't want to be hurt, she thought. I can't bear to be hurt. I don't want to lose an arm, or a leg, or burn. Please, don't let me be hurt!

Riendeau saw, out the window, the left wing tip nearing the ground. She waited for it to smack against the pavement, sending them into a cartwheel. Once more she shut her eyes and both hands clutched at the edges of her seat. Her fingernails tore into the fabric.

From somewhere at the front of the aircraft came a shout: *"Brace!"*

204

38

Gimli

Colin Nisbet sat in the cockpit of a Cessna 152 at the edge of runway 32 Right, preparing to take up a student. The two private flyers were in the midst of their final check when a voice interrupted. Through the crackle of radio static they heard, "Clear the area!"

"What was that?" the student asked.

"I don't know," Nisbet replied. He looked instinctively to the south, for if an aircraft was in trouble it was most likely to be approaching from Winnipeg. Then he saw the unmistakable profile of a 767 on final approach to what used to be runway 32 Left.

39

The Cockpit

Quintal was still searching through the Quick Reference Handbook for information on how to lock the forward landing gear in place, when Pearson's side-slip maneuver threw him against the inner edge of his seat. The Quick Reference Handbook slid from his grasp.

He glanced out the cockpit window. How much time did they have? They were close enough for him to see the black scuffmarks of tire burns, a sure sign of a well-used airport. But he saw something more.

There are people standing in the middle of the runway! he said to himself. And they can't hear us.

Pearson's attention was transfixed upon the threshold of the runway. It was home. He had to judge their rate of descent without instruments, and could not afford, even for an instant, to alter his gaze from the target point. Every ounce of his energy was sapped by his effort to maintain the side-slip. Perspiration flowed from his brow. He grimaced in unconscious pain, attempting to fly a precise glide slope that would bring the aircraft in at 180 knots as close as possible to the threshold. The computer in his own head made constant calculations; his hands on the yoke and his feet on the rudder pedals worked to increase and decrease the drag, as he normally would do with engine power. In response, the aircraft sped up or slowed down, dropped faster or slower. Each tiny movement by the pilot was reflected and magnified by the aircraft.

C'mon, baby, he said to himself. Right down the middle. Right on the numbers.

Pearson held the side-slip doggedly—for so long, in fact, that Quintal feared the left wing tip would hit the ground. The airplane was only about forty feet aloft when Pearson eased up on the controls.

The silent 767 leveled off and zeroed in on target.

Quintal realized that Pearson was unaware of the people on the runway. Should I tell him? No, it is too late. . . .

40

Gimli

Thirteen-year-old Cam Berglind pumped his legs hard against the pedals of his bicycle, trying to gain the advantage over his friends in the impromptu race. Like his buddies Art Zuke, fourteen, and Kerry Seabrook, eleven, Cam piloted his own sleek BMX racing bike. The chrome-plated beauty cost $1,500—an extravagant expense for many families, but these were the sons of auto racers; even their toys were built for speed.

Late afternoon at Gimli was always a highlight for the boys. The races were over for the day, the course was closed, and it provided the perfect playground. They had cycled to the south, away from the campers and vans parked adjacent to the straightaway and closer to the threshold of what was formerly runway 32 Left.

While pedaling furiously, Cam chanced to glance up, to the south, to the strange specter of a silent 767, canted at an odd angle, its left wing down. It was losing altitude rapidly. Transfixed, Cam saw it level off and point its nose directly at him. In that instant, he knew he was going to die.

"That pilot's crazy!" Kerry Seabrook yelled. "What's this jerk doing? He's landing on a drag strip!"

The three boys turned their bikes around and raced away from the onrushing invader.

Cam yelled out to the adults who clustered at the edge of the runway.

"Plane!" he screamed. "Crash!"

No one seemed to hear.

41

Flight 155

From the cockpit of Flight 155, soaring overhead at 35,000 feet, First Officer Gilles Sergerie strained for a glimpse of the stricken airliner as his own aircraft crossed over the edge of the cloud cover and into clear sky. He was an RCAF veteran, and knew the Gimli airport. With this as a visual reference, he now spotted a silvery dot against the landscape that must be Flight 143, carrying his friend and neighbor, Maurice Quintal. The aircraft was lined up on approach to the runway, which, from this altitude, appeared to be no more than an inch in length.

Sergerie continued to chain-smoke as he waited for whatever lay ahead. Gimli was off to his right and sliding past on the side. There was nothing that he, or anyone, could now do to help. The fate of the aircraft was in the hands of Bob Pearson and Maurice Quintal.

Sergerie felt helpless as he saw the tiny speck reach the threshold of the runway. And then he saw a mushroom cloud of white smoke rise up, engulf the aircraft, and screen it from view.

42

The Cockpit

At 0138 GMT, or 8:38 P.M., local time, twenty-nine minutes after the first indication of trouble, Flight 143 lurched heavily down upon the runway.

Normal touchdown point is 1,000 feet beyond the threshold. Pearson was almost precisely on target, accomplishing the incredible feat of landing a powerless jetliner 800 feet past the threshold. But there was not an instant to savor the victory. Two tires blew in the right main landing gear. Careering forward at 170 knots, far faster than normal, now he had to stop the craft before it slammed into something.

Pearson jammed the balls of his feet high up on the rudder pedals and pushed with his final reserve of strength to activate the brakes. The nose dropped. He anticipated the familiar thump of the forward gear touching down. Instead he heard what sounded like the explosive *bang*! of a 12-gauge shotgun fired at close range. The right engine nacelle scraped the ground. They were now sliding down the runway on their nose and an engine pod amid a cascade of sparks.

Lifting his eyes to look down the runway, Pearson caught a fleeting glimpse of a boy on a bicycle, and faced the most excruciating decision of his twenty-six-year career.

I am not going to hit people, he vowed. If he had to veer off the runway, he would do so.

43

Gimli

It was the kind of balmy summer afternoon that demanded relaxation. All along the final straightaway of the racecourse, the distinctive *pop*! of beer cans opening meshed with the giggles of children playing on the concrete strip formerly known as runway 32 Left.

David Glead and his wife Linda Jackson tended to their grilling steaks, enjoying their role as hosts to David's sister Jennifer and her husband Steve Barrow, visiting from Vancouver. David's shiny black 1970 Lotus Cortina emblazoned with a bright "95," sat next to their camper-van.

Off to his right, David heard the screech of tires, like the sound of a dragster burning rubber. He glanced over and saw the unexpected image of an airliner touching down. A keen fan of commercial aviation, David's immediate reaction was one of curiosity. He turned to Linda and said calmly, "You know, that's a brand-new 767. It's a beautifully engineered plane."

Then he did an incredulous double-take and curiosity gave way to fright. The aircraft seemed to have quadrupled in size. "My God! This is big!" he exclaimed.

* * *

Sergeant Bob Munro and Constable Roy Fenwick of the Royal Canadian Mounted Police were on routine patrol. Munro had driven the squad car out to the town of Fraserwood, ten miles west of Gimli, and was now headed back. The route took the two men along the two-lane country road immediately north of the airport. They were nearly adjacent to the runway when a call came in over their police radio, alerting them to the fact that a Boeing 767 was on its way to Gimli with no fuel and a load of "about 100 passengers."

Looking to their right they saw that the jet had just touched down. Its nose hit heavily against the pavement, creating a brilliant display of sparks and flowing smoke.

"For God's sake, don't roll over!" Munro shouted.

Cam Berglind approached the camper owned by Pat and Jo-Ann Barry. It was parked at the southernmost end of the racing zone because Pat Barry was the "scrutineer," the official who certified each car prior to racing.

Cam risked a quick look over his shoulder. The aircraft was on the runway now. He saw the nose bounce against the pavement. Smoke billowed out, engulfing the front.

"Plane! Crash!" Cam repeated. And now he added, "Fire!"

Pat Barry grabbed his sons. He scooped two-year-old Derek into his arms and grabbed five-year-old Danny by the hand. "Run, Jo-Ann!" he screamed to his wife.

Jo-Ann was in the camper, washing the supper dishes. She heard the warning above the sound of her radio and glanced out the window to see her husband and sons fleeing the racetrack. She turned to another window and saw others racing toward her, armed with fire extinguishers. Our camper is on fire! she thought.

Rushing outside, she realized that the men were running

216

past her, to her right. She glanced toward the edge of the runway and her eyes widened.

A cloud of smoke was approaching! Fast! Like a burning, supercharged dragster. It drew closer with alarming speed, and as it did so, Jo-Ann realized that it was not a car at all. It appeared to be a burning airplane.

44

The Cabin

Susan Jewett felt the aircraft smack onto the ground, bounce, hit once more, and hurtle forward at breakneck speed. This is it, this is good-bye. . . . she repeated to herself. Oh, my God, Victoria . . . how will my husband ever manage a one-year-old by himself? Oh, Victoria, I wanted to see you grow up. . . .I wanted you to have brothers and sisters. . . .

What is ahead of us? What is it that we will crash into? What will it feel like to be shattered into a thousand pieces? Nigel Field knew these to be his final thoughts.

Bob Howitt, hunched forward in his seat to form a protective cocoon around three-month-old Katie, waited for the concussion of impact. He felt the nose of the airplane dip forward and was assaulted with the pungent odor of metal scraping against rock.

Richard Elaschuk did not keep his head down, as instructed. He sat upright with his left arm firmly on the

219

back of his two-year-old son Stephen. In this position he could look out the window and see that they had touched down on some surface that resembled a runway.

Pat Mohr felt the body of the craft shake, threatening to burst apart. Her body tensed. Her hand was welded to Heather's as she waited for darkness.

45

The Cockpit

Without a nosewheel to steer the aircraft, Pearson used differential braking, alternating the pressure of his feet against the left and right brake pedals in order to keep the aircraft in the center of the runway. He now saw more people scattering, running off to his left, away from the runway. *Hold it, hold it, hold it,* he commanded himself, gauging the approaching moment when he might have to steer hard right, off the relative safety of the pavement.

A new image appeared, improbable and confusing. A low metal guardrail stood in the middle of the runway, set along its length. Pearson leaned heavier on the right brake. The aircraft veered only slightly, skidding. The left side of its nose glanced off the low metal fence, shearing off the round wooden posts at their bases.

46

The Cabin

Sweat pouring from his forehead, Bryce Bell anticipated the massive, final bang, the fireball of pain and annihilation, the ultimate smack, he thought. Sharp cracks echoed through the cabin. Telephone poles, my God, we're hitting telephone poles!

47

Gimli

Watching from the side of the runway, David Glead stood welded in place, transfixed. The airplane was approaching so fast. He was sure there was no way the pilot could stop it before it smashed into the race cars, barbecues and people. Some ran toward safety, away from the runway. Others, like Glead, were frozen.

He saw a cloud of smoke billow from the 767's nose as it crushed the wooden posts of the guardrail, and he saw and heard the lumbering aircraft slow and then, amazingly, stop short.

Now there was silence. In the middle of the runway lay a motionless aircraft, its nose to the ground, surrounded by a growing cloud of billowing smoke.

48

The Cabin

The body of the plane was intact. They were on the ground, and as the stunned passengers lifted their heads and peered about, the realization that they were alive resulted in a surge of euphoria.

A cry of "Yahoo!!!" echoed throughout the plane. Spontaneous applause erupted as the individuals aboard Flight 143 comprehended the incredible fact that they were, indeed, still among the living.

Flight attendants snapped into action. They were on the ground, yes, but the danger of fire or explosion—or both—was still very real. "Come on, come on, we've got to get out of here!" they shouted.

49

The Cockpit

Pearson and Quintal had only a moment to reflect upon the fact that they were safely on the ground.

Smoke from an unknown source now poured into the cockpit. *Fire!* Pearson thought. To an airplane pilot there is no greater emergency.

Despite the evidence of fire, the pilots rushed through a final emergency checklist. Quintal deactivated the fuel control switches to guard against an explosion, realizing as he did so the absurdity of *that* action. There was no fuel!

Pearson pulled on the parking brake, which was also unnecessary. Both pilots pulled fire control switches to shut down the oil flow and the electric and hydraulic systems. They switched off the battery, and checked to reassure that all systems were deactivated.

By now, the smoke in the cockpit was so thick they could barely see and almost could not breathe.

50

Gimli

Cam Berglind, still racing his bicycle away from danger, scurried around the side of a small service building just off the western edge of the runway. Here, he encountered a group of twenty or thirty members of the Winnipeg Sports Car Club, who were discussing the results of the fun rally. The building had blocked their view of the crash, and, amazingly, no one had heard a thing.

"A plane crashed! A plane crashed!" Berglind shouted.

The boy has a vivid imagination, Kevin Lloyd told himself. Nevertheless, Lloyd strode to the corner of the building and peered around, expecting to see a small Cessna that had, perhaps, ground-looped. Instead he saw a 159-foot-long 767, nose down, exuding thick, black, oily smoke.

Lloyd ran to his car and searched frantically for his emergency equipment. "Where's the first-aid kit?" he shouted in frustration.

Two friends heard the desperate question and brought

the kit. They had borrowed it to treat their Old English sheepdog, who had scraped the pads of his paws while romping on the rough surface of the tarmac.

Lloyd grabbed the kit and ran for the airplane, passing a man who was running away from him, screaming, "It's going to blow up!"

"Get fire extinguishers!" someone yelled.

"Stand back!" Keith Berglind ordered his son. Then he ran off with the others to help.

David Glead and other club members grabbed for the extinguishers that were placed at strategic intervals along the racetrack. Holding one of them in his lap, he vaulted onto the front fender of a car as it sped off toward the wounded airplane. Smoke now billowed freely from the nose.

As they neared the 767, Glead saw the first of the emergency chutes deploy, and he knew the crew was attempting to evacuate the passengers—fast!

Am I dead? wondered Danielle Riendeau. No, no we are down, we are down without engines. We are alive. We are all really alive! She watched herself, as if from a lofty height, unbuckle her seat belt and walk briskly toward the left over-wing emergency exit.

A small young wisp of a girl was sitting there, next to the window, immobile, despite the fact that the cabin was filling with smoke.

"Open the window," Riendeau shouted to the girl.

The youngster tried. It was too heavy. She couldn't budge it.

Reaching across her, toward the emergency exit, Riendeau found a strength she did not know she possessed. She rammed her shoulder against the door and forced it open. Then she stepped aside to let passengers flee past

232

her, out the door, across the wing, and down the slide to safety.

With deliberate speed that fell short of panic, Michel Dorais led Shauna Ohe to Riendeau's exit over the left wing. Seconds seemed like hours as the smoke thickened.

Reaching the door, Ohe peered beneath her and froze. She was supposed to step out onto the wing, then jump onto an evacuation chute. Here in the center of the aircraft the slide seemed to be a precipitous distance from the ground. Dorais pushed her out onto the wing. As she mustered her courage and jumped onto the slide, she realized that Michel had remained behind in the perilous environment of the passenger cabin.

Ohe flew down the slide and bounced onto the tarmac, landing on her backside. She stumbled to her feet and staggered as pains shot through her lower back. Most terrifying was the knowledge that Michel was still on the airplane.

In one quick burst of movement, Nigel Field unbuckled his seat belt and stood up. His hand rested on the hatch, over the right wing, poised. "Don't open it," Desjardins' voice commanded.

For some reason he could not fathom, passengers ignored this exit. Some took the opposite route, over the left wing, but most scurried uphill toward the rear of the airplane. Field was aware of a sobbing sound from somewhere in the rear—not the high-pitched wails of an infant, but the deep, heavy sobs of an adult. At his side he heard a flight attendant say to him, "We would like you to go and help the people get through this. . . ." She was looking and gesturing toward the front of the aircraft. Only then did he become aware that the cabin was filling with smoke. That was why everyone was moving to the back, away from whatever was burning. The specter of Cincinnati hung in the clouded air of the passenger cabin.

Field ran forward, toward the smoke, to the tiny first-class cabin. Then he moved farther forward to the right-side front door. It was open, its emergency chute inflated and angled gently toward the ground.

A middle-aged woman stood adjacent to the door, anxious and confused. Field helped her onto the chute and then turned back to the first-class cabin to double-check. Reassuring himself that it was empty he felt an unaccountably comforting sense of isolation. Here he was, alone on a burning aircraft, yet he felt no compulsion to leave.

He took one final look around and then slowly, meticulously, walked to the front starboard exit and stepped off onto the ground.

When Rick Dion saw the smoke, he thought immediately, It's time to move our butts. "Let's get on with it," he said.

Cradling three-year-old Chris in his arms, Rick guided Pearl toward the left over-wing exit. Others were already evacuating.

"I'll take Chris down with me," he said to his wife. "When you get down there, run away from the plane."

They reached the exit with Pearl in front. "Go ahead, go ahead!" Rick commanded.

Pearl looked down at the frightening drop. "I can't," she cried. "I can't!"

"I'll go down with Chris," Rick said. "There's nothing to it."

With his arms wrapped around about his son, he hopped onto the slide and sped downward. He hit the pavement feet first and tumbled forward, head over heels, absorbing the impact with his own body, but shielding Chris.

In the aircraft, Pearl still balked. Suddenly someone shoved her, and she slid to safety. Together the Dions raced across the runway, away from danger.

The instant the airplane stopped, Mike Lord unbuckled his seat belt and jumped into the aisle at the left rear of the plane, following his instructions to block the exit until Annie Swift had opened the door and released the emergency chute. White-faced passengers moved rapidly toward his exit.

"Okay now, Mike!" Swift called. Lord turned and stared out the open door, amazed at how far away the ground appeared This was like jumping off the roof of a three-story house!

There was no time to argue. For all he knew, the plane could explode in an instant. Crossing his arms in front of him, so that the harsh surface of the slide would not burn his bare skin, Lord jumped into space. He hit the chute in a sitting position and picked up speed so quickly that he instinctively grabbed the edge. Friction tore patches of skin off the palms of his hands.

Here at the rear of the plane, with the tail sitting high because of the collapsed nose gear, the chute ended about three feet above the runway. Lord crashed onto the pavement, slicing off the bottoms of his socks. He pitched forward, somersaulting ahead more than ten feet from the end of the chute. Picking himself up, now unconcerned about his own well-being, he ran back to the base of the chute to assist others as they came down.

"Come on, come on . . . everybody off!" Susan Jewett heard the volume of her own voice escalate to a near-shout.

235

When she opened the emergency exit at the right rear of the aircraft and activated the inflatable emergency slide, unaware of the nose-down attitude of the airplane, she was incensed at what appeared to be inadequate design. The bottom of the chute hung limp, ending several feet short of the ground. They are crazy if they think people are going down there. This is ridiculous, she thought, livid that her passengers had defied death in the sky only to be confronted with an evacuation procedure that could very well kill them on the ground.

Her able-bodied passenger was the first one down her chute. The moment he bounced onto the pavement he hopped to his feet and ran for safety.

"Come back!" Jewett called from thirty feet above. "You have to help the others."

He heard her plea, remembered his job, and took a post at the base of the chute.

An elderly woman sat alone, a few rows ahead of Ken Mohr. For a moment he thought he should help her but his own family was his priority. He grabbed Crystal's hand and headed for the exit behind him, feeling a stab of guilt for leaving the elderly woman behind.

As father and daughter stood at the top of the tail section, looking at the chute that would spirit them to the ground below, Crystal became terrified. "No, I don't want to go," she cried.

"Come on, we're going," her father ordered. He grabbed the frightened youngster, wedged her tiny body between his knees, and they slid down the chute, tumbling onto the ground. The surface of the tarmac ripped through the seat of his pants.

On the other side of the aircraft, Pat and Heather Mohr

faced the same steep angle. Like her sister, Heather resisted.

"Hurry, come on," the flight attendants implored.

Without her glasses, Pat could barely see the angle at which the chute careened off the side of the plane, but to Heather it seemed as if they were about to leap off of a skyscraper. She pulled her mother aside to let a few other passengers go ahead. The eleven-year-old girl cried and shivered, although the evening air was warm.

"Come on!" Pat Mohr commanded. "We're going *now!*" Clutching her daughter's hand, Pat forced Heather onto the top of the chute. They slid rapidly to the bottom, side by side. Heather landed on the foot of the passenger in front of her. Pat, attempting to get out of the way quickly, rolled into a backward somersault, leaving her dazed and disoriented.

Peering underneath the tail of the airplane, Ken Mohr saw his wife and daughter tumble onto the tarmac. Dragging Crystal by the arm, he ran around the rear of the plane and gathered his family. He threw his arms around his wife and, for the first time since the nightmare had begun, Pat Mohr gave in to sobs of relief.

Lillian Fournier and Pearl Dayment looked at one another in astonishment that they were not only alive but in one piece. Their reverie was short-lived.

"We're not safe yet," warned Susan Jewett. "Come quickly!"

Lillian looped her purse around her neck and grabbed her shoes. She had difficulty navigating through the aisle without her glasses. Pearl helped. She reached the emergency exit at the right rear of the aircraft and, squinting, tried to focus on the chute. She had always been frightened of heights, and now the steep angle and the altitude of the chute terrified her.

"No," she said, "I'm not going."

"Get on," Jewett prodded. "It's better to break an arm than to blow up!"

"No," Lillian insisted. "I won't."

"Get on!" Jewett repeated. "Now!"

The insistence in the young woman's voice broke Lillian's resistance. Lillian did as she was told, sitting down at the mouth of the chute. Elbows bent, arms up, she slid down the steep angle, her skirt flying. She slid off at the bottom, feet up, head down, and cracked her skull against the pavement. Her world went black.

Two men ran to her side, struggled to pick her up and steadied her as she gained a fuzzy consciousness.

Above them, it was Pearl's turn. "You can't take those with you," Jewett said, pointing to Pearl's shoes and purse. The woman would need her hands free in order to balance herself on the chute.

If I'm going to be killed, I won't need the money, Pearl reasoned. She stowed the handbag and shoes on the floor, next to the emergency exit, clambered onto the chute, and flew downward. By the time she reached the bottom, her arms were raw and bleeding.

Joining her dazed sister, Pearl scurried away as fast as she could. After a time she paused, turned, and stared at the huge wounded plane she had just fled. Thick smoke billowed where the nose gear should have been.

She said a prayer of thanks.

After pushing Shauna Ohe out the doorway of the left over-wing exit, Michel Dorais stumbled forward through the smoky, darkened cabin. Carefully he checked each seat to make sure no one was left behind. Few passengers had been seated forward, and they had obviously evacuated quickly.

By the time Dorais reached the left forward door the cabin was bathed in so much smoke that he could barely see. Realizing that he was at the door and seeing the edge of the emergency chute, he jumped, expecting to slide down to the tarmac. His body was stunned by a sudden jolt. Here, up front, the fuselage was sitting upon the surface of the pavement.

He peered back and beneath the left wing. Ohe still stood near the base of the slide, reeling from the blow to her back. He raced toward her, and they embraced.

Michel is a hero! Ohe thought. He stayed behind to help the others.

"Quick," Dorais said. "We've got to get away from here."

Pauline Elaschuk had jumped up immediately, cradling four-month-old Matthew in her arms, ready to race for the emergency exit at the rear of the plane.

"Wait for the ramp to be inflated," Richard said.

The aisle was filled with passengers awaiting their turn to exit. As smoke curled in from the direction of the cockpit, Pauline and Richard Elaschuk remembered Cincinnati.

After Richard had unbuckled two-year-old Stephen, the family joined the swiftly moving queue. Ahead of them, Pauline watched as an older woman slid down the chute and tumbled face-forward onto the tarmac with a sickening *smack*!

My God! I'm going to kill my child, Pauline realized. Until this moment, aware that she had no means of exerting control over the situation, she had been resigned to the dictates of fate. Now, standing at the top of the emergency slide, peering down some thirty feet, she had a vision

239

of hitting the pavement, crushing Matthew between her body and the ground.

Others were behind her now, yelling at her to go. Smoke was pouring in toward the rear. I'm supposed to jump down here and kill my baby? she thought.

The first few passengers had arrived at the bottom of the chute with such speed that Mike Lord was unable to prevent them from tumbling to the pavement as he had done. He tried a different approach. Standing to one side, he attempted to hook passengers under the armpits. This way he could break the fall a bit but, he muttered to himself, "They're still landing on their asses."

He heard a woman above him, screaming. "You catch my baby!" He looked up to see Pauline Elaschuk standing in the door of the plane, clutching Matthew to her breast. "You catch my baby!" she repeated in a threatening tone.

"Jump. I'll get him," Lord promised.

Down the slide they came, Pauline holding her body taut. Lord lunged to grab her under the armpits and managed to break her fall enough so that she did not hit the pavement too hard. She landed on her buttocks and her clenched teeth cracked together, but she clutched Matthew safely in front of her. Lord helped her to her feet and turned his attention back to the chute.

Looking around her, Pauline Elaschuk saw a panorama of confusion. Above and behind her was the huge aircraft, smoke billowing from its forward end. Despite the obvious danger of explosion, passengers were milling about. You fools, she thought. Run! But she, too, remained in place. She would not leave without Richard and Stephen.

From the top of the chute, Richard yelled to Mike Lord, "I have a kid, too!"

"Okay," Lord yelled back.

What a stupid slide, Richard Elaschuk thought. It goes almost straight down.

Richard jumped onto the chute with Stephen held tightly in his arms. They barreled into Mike Lord, who absorbed the blow like a linebacker, but somehow managed to keep his balance. Richard fell to his left, cracking his knee and elbow hard against the pavement, but buffering Stephen successfully.

He scrambled to his feet, grateful to see Pauline and Michael safe. Instinctively the family ran away from danger. Pauline's feet were covered only with shredded nylons, but she felt no pain.

Richard glanced over his shoulder, back at the airplane. The engine nacelle on this side was only two feet off the ground. "How did that engine not hit the ground?" he muttered. "How did we not go into a cartwheel?"

His questions unanswered, he continued to run at Pauline's side.

From the far end of the runway, a stranger appeared heading toward the aircraft. "Can I help you?" he asked. "Can I take your baby?"

Pauline's lungs ached from the exertion of her sprint. "No," she panted. 'I'm all right. Someone else may need you."

The Elaschuks ran on, stumbling toward a van where sympathetic hands pulled them to safety.

Finally Pauline felt her body begin to quiver uncontrollably. Now that they were safe, the tears would not be denied.

Like a nervous mother anxiously counting the fingers and toes of her newborn infant, Bryce Bell checked himself thoroughly. He could not quite believe that he was in one piece. He had been prepared for a fiery blast into eternity and instead found himself still alive. Adrenaline swept through him like a brushfire.

He bolted from his seat and grabbed the nine-year-old girl who had been put in his charge and headed for the exit at the rear of the plane. He watched as the girl's mother, with her younger child on her lap, careened down the chute. She fell back, hit her head soundly on the pavement and bounced, rather ungracefully, on her backside. Convinced that the plane would momentarily explode, Bell held his young charge tightly around the waist and positioned himself at the mouth of the chute. They slid quickly to the bottom, the weight of the youngster pushing his elbows down onto the chute, burning off patches of flesh. Bell landed solidly on his derriere and let out a resounding shriek, just as the child's mother grabbed her out of his hold and shepherded her to safety.

Bell picked himself up and turned to assist the other passengers. A woman, perhaps in her eighties, slid toward him. He reached for her just as her feet hit the tarmac. He missed, and her head fell back against the pavement. Her eyes rolled back in her head, revealing pockets of ghostly white. Bell did not know if she was unconscious or dead. "Christ," he muttered, hoisting her up onto one massive shoulder and running away from the airplane. He reached a grassy area at one side of the runway and gently laid her down.

"Stop, come back! We want all the passengers over here," Susan Jewett yelled to him.

"What the hell?" Bell asked no one in particular. He ran back to the still-dazed woman, picked her up once again, and headed toward an assembling group of passengers and crew. She seemed to grow heavier by the moment. Bell felt as though his hemorrhoids had settled in the vicinity of his ankles.

Turning the woman over to the care of others, he looked around, attempting to orient himself.

From above, Jewett eyed him with admiration. During the flight he had been the ornery passenger, the one to give her trouble. Now he was the one who had come to the aid of the wounded woman.

As the Mohrs enjoyed a moment of relief, standing near the bottom of the slide at the rear of the airplane, a flight attendant approached them and shouted, "We have to keep going. Let's not hang around the back of the plane." She offered to carry Crystal, but Pat Mohr shrugged her off and hugged her daughter close.

"Come on, let's get out of here," Ken Mohr said.

The family ran away from the airplane until they encountered a fence. "What the hell is this fence doing here?" Ken wondered aloud. "They don't have chain link fences on runways."

Then he saw a man run past him, toward the airplane, carrying a small, hand-held fire extinguisher. What is this Mickey Mouse stuff? he asked himself. They can't fight an airplane fire with that thing! Now he had his first chance to catch his breath, to look around. "This isn't Winnipeg," he said to Pat. "Where are we?"

He heard someone mention Gimli, and he tried to place it in his mind. He had heard of a wine called Gimli Goose that cost about $2.98 a bottle, but he didn't know of a town called Gimli.

His confusion gave way to euphoria, and he thought of the skill of the pilot who had brought them down safely in—wherever it was. We are alive and it's because of him, he thought. His eyes searched for the pilot; he wanted to shake his hand.

* * *

243

Joanne and Bob Howitt had remained in their seats as others scampered for the emergency exits. They did not want to get caught in a crush with two small children.

When the aisles on either side of them had cleared, they moved rearward. Joanne went down the left aisle, tugging three-year-old Brodie by the hand. Bob took the right, clutching baby Katie, still in her harness, to his chest.

When they reached the left rear emergency door, Joanne and Brodie hesitated. The slide was angled almost straight down. The height was petrifying. Joanne saw a woman fall off the edge of the slide and sprawl upon the pavement below. "Oh my God!" Joanne exclaimed.

"I'm scared!" Brodie wailed.

His mother tried to calm him. "It's like going off a slide, Brodie," she said. "I'll hold you. I promise."

Joanne scrambled to a sitting position, pulling Brodie onto her lap, and slid. Mother and son flew down the chute and crashed to the ground. Brodie flew free from his mother's grasp, bashed his head against the pavement, and burst into tears.

Joanne rushed to her child. The tip of his nose was scraped and he was bleeding slightly from a few scratches, but he appeared to be all right.

"Run!" a man said to her.

Joanne looked up to see the man she had christened the Preacher—the Asian who had remained immersed in his holy book during the previous twenty minutes of terror. He gestured fiercely. They must get away from the airplane.

"My husband is still on the plane," Joanne shouted.

Without hesitation, the Preacher scooped Brodie into his arms and ran for safety. Joanne looked around her, then back toward the emergency exit, so high above her, searching for her husband and baby.

I'm not going down this ramp, Bob thought. It's too dangerous. He turned to the opposite side of the airplane, crossed the galley, and stood at the left side door where Joanne and Brodie had exited moments earlier. His mind raced. This is not good, he said to himself. If I go down this ramp I will tumble onto my face and land on top of Katie.

"Jump! Jump!" ordered a flight attendant. "Then get the hell away from the aircraft. Run as soon as you get on the ground."

"Just a minute. I've got a baby here."

The flight attendant grabbed Bob's hand and they sat side-by-side on the twin slides of the emergency chute. In the few seconds it took to hurtle thirty feet down, Bob commanded himself, Land on your ass. Throw your feet forward and land on your ass. Don't fall on the baby.

As he neared the bottom of the slide he kicked his feet high out in front of him. Then, just as his body left the slide and flew into the air, Mike Lord stepped out, seemingly from nowhere, caught Bob with one arm in the crotch, threw another arm behind his back, and deposited him and baby Katie safely onto the tarmac.

Holy shit, Bob thought. I'm on the ground.

Now only a flight attendant remained above Mike Lord at the rear of the aircraft. As she jumped onto the chute, Lord lunged directly into her path, holding his arms out to catch her. Halfway down the slide, she spread her legs apart, slamming into Lord, one leg on either side of him. He clutched his arms around her, staggered backward, and fell to the pavement, cracking the back of his head.

Lying on top of the muscular bachelor, the flight attendant suddenly smiled. She looked into his eyes and said, "Nicest arms that ever held me. Let's go!"

Pearson and Quintal finished their checklist and fled from the cockpit, coughing from the heavy smoke. "Get me the extinguisher!" Pearson yelled to Quintal.

As the first officer searched in the smoke-induced darkness for the fire extinguisher, Pearson saw a woman struggle down the aisle, laboring to walk down the steep forward slope. It was the first moment he realized that the aircraft was nosed down so severely. The woman left via the forward emergency exit.

Pearson and Desjardins, flashlights in hand, made sure that everyone was off. Desjardins then exited down one of the rear emergency chutes, but Pearson returned to the forward section. "Where is the extinguisher?" he asked Quintal.

"It's not there," Quintal answered. "Maybe it disengaged on landing."

Satisfied that all the passengers were off, the captain's concern was now for his airplane. He wanted to locate the source of the smoke and squelch the fire before it did any more damage. He and Quintal left Aircraft 604 from a forward exit, so low to the ground that they simply stepped off onto the runway.

Rick and Pearl Dion sat on the tarmac with their son Chris, recovering. Pearl was in shock when Rick confessed an earlier lie. He said, "I didn't want to tell you on the plane, but we had no fuel at all."

"I don't believe you," Pearl said.

Once he was certain that the passengers and crew were safely away from the aircraft, Captain Pearson went back on board to locate a fire extinguisher from the flight deck.

246

The density of the smoke made it possible to see. He took a CO_2 extinguisher from the flight attendant position, exited the plane and began spraying the nose area. He was joined by several members of the Winnipeg Sports Car Club, who offered their assistance and a fresh supply of fire extinguishers.

The men fired several bottles into the nose area. Pearson boarded the plane once again, but the smoke was still too thick. He could not see. He could not breathe.

Pearson left the aircraft, grabbed a fire extinguisher and tried again. This time he was accompanied by Sports Car Club volunteers. Together they sprayed additional CO_2 bottles around the area of the rudder pedals, where the smoke seemed to be originating. Finally, the smoke began to dissipate.

The Gimli Fire Department arrived and located the smoke source. Under the belly of the aircraft, insulation was burning softly, apparently ignited by the friction of the nose-down landing. Firemen extinguished this, and a gentle breeze carried off the remaining wisps of smoke.

Passengers milled about, some euphoric, some angry, all disoriented. The old woman whom Bryce Bell had carried now lay motionless upon a stretcher. A young woman sobbed loudly for a few moments, then stumbled around the tarmac in her stocking feet, dazed and numb. One man worried about the frozen lobster he was carrying home from Prince Edward Island; he had left it on the plane.

David Glead of the Winnipeg Sports Car Club heard someone say, "This can't be Winnipeg," and he thought: Boy, are you in the wrong place!

Mike Lord sat on a small fence, catching his breath, as Cam Berglind jabbered excitedly into his ear. "Me and my

247

friend were racing our bikes," he said, "right down this runway. We looked up, and suddenly the whole sky was Air Canada!"

The normally loquacious Lord was speechless. What he wanted most now was a cigarette. He pulled a half-filled pack out of his breast pocket, along with the lighter that he had kept there, contrary to safety instructions. As he lit up, he found himself suddenly surrounded. Everyone, it seemed, wanted to bum a cigarette.

Annie Swift only smoked when she was nervous or excited, and she puffed away furiously now. Gazing at the unreal sight in front of her, she suddenly felt an aching sensation throughout her body. She wondered if she was sore from the exertion of the past few minutes, or from horseback riding this morning.

Her colleague, Danielle Riendeau, was in pain also. Her shoulder ached from when she had thrown it against the emergency exit.

As the two women hugged in relief and cried out their tears of exhaustion, an old woman approached, the first-time fearful flyer whom Riendeau had attempted to console. The woman embraced Reindeau and cried, in French, "Thank you for saving my life!"

* * *

Joanne and Bob Howitt, reunited with each other, found respite at the far end of the runway, where a woman member of the Winnipeg Sports Car Club offered to drive them into town in her Datsun.

Joanne looked about her, noticing for the first time that they were not at a large municipal airport. What was a Coca-Cola stand doing in the middle of the runway? Suddenly she was very angry. "This isn't Winnipeg!" she screamed.

248

"Well," Bob said, "wherever it is, we're on the ground, eh?"

Another passenger approached Bryce Bell and said, "I wonder if this will interfere with me catching my connecting flight."

Bell stared at the man incredulously.

Cam Berglind, having escaped the path of the crashing airplane, stood off to one side, watching his father and others help evacuate the passengers.

Maybe I should phone the radio station, he thought. I could get some money for a news tip.

Sergeant Bob Munro of the Royal Canadian Mounted Police cordoned off the area to keep the curious away from the quiet, empty aircraft. By now, he knew, investigators from the Ministry of Transport were on their way There would be many difficult questions to answer. How in the world could a regularly scheduled airliner run out of fuel in the midst of flight?

Equally incredible, but on the positive side, was the feat accomplished by the pilots. They had performed the impossible. When Munro learned that Pearson had put the powerless jet into a side-slip in order to avoid overshooting the runway, he remarked, "That's almost impossible!"

Michel Dorais and Shauna Ohe were catching their breath when Annie Swift ran up to them. Impulsively, Swift caught Ohe in a bear hug and cried, "We've just witnessed a miracle!"

Dorais sought a more logical explanation for the events of the past twenty-nine minutes. He considered the data: Takeoff had been delayed because of a problem with the

249

fuel gauges. They first planned to divert to Winnipeg because they had to fix the fuel gauges. Suddenly they had headed for Gimli with, quite obviously, no leeway for maneuvering. There had been some kind of fire. Why was the stricken aircraft not incinerated in a holocaust? It added up to one logical conclusion.

As Swift ran off to talk with other passengers, Dorais said to Ohe, "I'll bet those bastards ran out of fuel."

Ohe said, "N-o-oh. . . . You've got to be kidding. That's impossible!"

Epilogue

On Tuesday, July 26, 1983, Diane Rocheleau, a mechanical failure analyst employed by Canada's Bureau of Aviation Safety, removed the fuel quantity processor from Flight 143 and hand-carried it to a Minneapolis testing facility operated by Honeywell Laboratories, the manufacturer of the unit. Together with a half dozen Honeywell experts, she conducted extensive tests.

In the laboratory, the investigators powered up the fuel quantity processor to see what would happen. After about two minutes it began to emit smoke.

Probing further into the tiny innards of the microprocessor, the team of investigators determined that the fault lay somewhere within the largest of six inductors encapsulated in an epoxy compound. Digging inside channel 2, they found the cold-soldered joint, the partial, improper connection that was the first link in the bizarre chain of events that produced the strange odyssey of Flight 143.

At Gimli, Aircraft 604 was repaired sufficiently to allow Air Canada's chief 767 flying instructor, Dave Walker, along with Captain Bob Clarke, to fly it to Winnipeg where it underwent further extensive repairs for more than a month.

It is still in service today, known to insiders as "The Gimli Glider."

Within a week of the accident, investigators from the Ministry of Transport assembled the flight attendants from Flight 143 in a conference room at Montreal's Dorval Airport as part of their routine investigation. The room featured large windows, fronting on the airport runway. As the flight attendants, one by one, related their accounts of Flight 143, they noted increasing activity outside. Military officials gathered to watch the demonstration of a highly touted new bomber. It took off and flew several passes over the airport.

"My God! He's going to hit the window," Susan Jewett exclaimed at one point, when the bomber appeared to be flying directly toward the conference room.

The aircraft leaped up over their heads and circled for another pass over Dorval as the investigators and flight attendants attempted to concentrate upon their business. Suddenly they heard the unmistakable, sickening roar of a crash. They looked up to see thick black smoke pouring from the far end of the runway. The pilots were rescued, injured but alive; the aircraft was lost.

This ended the meeting.

Minutes later, as Swift made her way through the airport on her way home, she chanced to encounter Don Cameron, the pilot of the Air Canada DC-9 that had caught fire in the air a month earlier and crash-landed in Cincinnati. Swift did not know Cameron, but she recognized him from the press coverage.

"I've just been on the Gimli flight!" she burst out.

Instinctively, the two hugged.

252

Lillian Fournier and Pearl Dayment returned to their lives in Pembroke, Ontario, shaken by their brush with death. For Lillian, the immediate problem was the debilitating effect of a head injury, incurred when she hit the tarmac at the base of the emergency chute. For two months she was unable to work a regular schedule at Miranucki Lodge. She sued Air Canada for compensation and settled out of court for $5,000.

A more lingering effect was a deep empathy for the victims of any airline disaster. Today, if she hears such a report on the news, she melts into tears. "You think of the terror, and you just hope it was fast," she says. "The horrible part is waiting for the unknown."

She adds: "Certain things make me cynical. We could all have been killed for such a stupid error. I don't think I'm going to fly anymore."

Sitting in the comfortable living room of her home in Richmond, Ontario, sipping coffee, Pat Mohr contends, "I'm not so apt to do things now that I don't want to do. We almost died. This is a second go-around and we've realized that we might as well enjoy life while we have the chance. I think we all enjoy the day more."

Her husband Ken adds with an ingratiating grin that the experience taught him to relax. In the past, he often succumbed to the pressures of his job at Siltronics, Ltd., where he oversees the manufacture of integrated circuits. He says: "Now I sit there and think, We've got this problem. Hell, we can handle this. You want to hear about problems . . . ?"

Heather and Crystal Mohr both own T-shirts with the inscription:

I RAN OUT OF GAS ON A 767

* * *

Shauna Ohe and Michel Dorais were married in Edmonton on June 30, 1984. The incident has changed them little outwardly, but Shauna Dorais says: "Sometimes you wonder how you would react if you were put into a life-or-death situation. Now we know." Remembering that Michel forced her out of the emergency exit and remained behind to make sure others got out, she says with pride and love, "My husband is a hero."

Weeks after the accident, after returning to her home in Alberta, Pauline Elaschuk began to experience severe and painful dental problems. Abscesses developed in two teeth and a third was cracked. Surgery was necessary—she lost a portion of her jawbone—and four years of painful treatment followed. Specialists confirmed that the injuries were in all likelihood due to the trauma suffered when she tumbled down the evacuation slide, concerned for the safety of four-year-old Matthew who was clutched in her arms. Air Canada officials sympathized with her plight and acknowledged responsibility, but refused financial liability. Pauline was unable to recover any damages, because she failed to file suit within 100 days of the incident, as required by Canadian law.

The Elaschuks still fly, but only when absolutely necessary. "Thinking about it," says Pauline, "I still have a brush with tears."

Bob Howitt found, in the experience, motivation to focus his energies more firmly upon his goals. He enjoyed his work as a soil scientist for the Alberta Research Council, and still does, but after July 23, 1983, he determined to pursue the long-dormant dream of continuing education. He enrolled as a doctoral candidate at the University of

254

Alberta. As of this writing he has finished his coursework and is laboring over his thesis.

Joanne Howitt, on the other hand, moved on to several different jobs since the incident, unable to find contentment until she realized that her priorities had shifted. When she redirected her attention away from the workaday world and toward Bob and her children, Brodie and Katie, she knew that she had learned to appreciate the most important things in life.

"I was incredulous beyond belief that an airplane like that could run out of fuel," declares Nigel Field. as he sits in the library of his home in Cornwall, Ontario. "I was very angry, and I had very mixed feelings about the crew. I realized that the pilot had done a fantastic job in bringing us down, but I also realized that he had some responsibility for getting us into that mess in the first place."

As manager of plant capacity planning for Canadian National Rail, Field still embarks upon frequent business trips, displaying the Englishman's legendary unflappable poise. "It's nice to know that in a situation where you are facing death—that you can face it with equanimity," he says. "I felt good about that afterward. To me that is fairly significant, because one doesn't really know how one will react until the time actually comes."

"I would be thrilled if I thought I never had to get onto an airplane again," Bryce Bell proclaims, and in the aftermath of the incident he made his sentiments well known. "In retrospect, I guess I was a bit of a shit," he admits. "I wanted to get even with the airline; I was so mad at them."

Bell took his story of terror and official malfeasance to the newspapers which, for a time, created a public apprehension over the airworthiness of the Boeing 767. He and

255

Bob Howitt attempted to organize the passengers as complainants in a class-action lawsuit against Air Canada, but the effort fizzled due to the intricacies of Canadian law. Nevertheless, they helped keep the issue in front of the politicians, which was a factor in the government's decisions to initiate a major federal inquiry. Finally Bell managed a personal metamorphosis.

"The incident at Gimli was the impetus that finally gave me the courage to change my life," he proclaims. "I worked for the government for sixteen years and I hated it. I found that everything else diminishes when your life is put on the line. So what if you give up your government job? So what if you have to scramble for a living?"

Bell quit his post as director of program services for the Alberta Department of Advanced Education, sold his house in Edmonton, and moved with his wife Margo and son Jonathan back to Larrimac, north of Ottawa. Taking up residence in the forest retreat originally built by his father, Bell launched a new career, designing and building rustic homes featuring deluxe handcrafted woodwork.

In 1987, Mike Lord, the computer operator from Montreal who manned the emergency chute at the left rear of the airplane, married his longtime girlfriend Danielle. His bride is wary of flying, but Lord has managed to focus his own apprehensions on Air Canada, rather than upon airlines in general. He recalls a recent experience: "I went to get a ticket from another airline, but it was too much money, so I walked over to the Air Canada ticket counter and looked around. I just stood there for a few minutes, wondering if I should buy a ticket from Air Canada. Then I said to myself, 'Nope!' Eh?"

Air Canada mechanic Rick Dion was transferred to a post at Vancouver Airport. The Dions live today less than a half mile from the U.S. border, which makes it convenient for Pearl to drive across so that she can buy milk in gallons, rather than liters.

"It was like a lot of other accidents," Rick details. "There were numerous factors involved. Where do you place the blame? You can start with the federal government and finish with the captain and point a finger at everyone in between. There should have been built-in safety factors to avoid it.

"On the other hand, once we were in the air and experiencing problems, lots of good things happened. That configuration of everything going wrong and everything going right probably won't happen again."

Pearl Dion agrees. Although she is still a white-knuckle flyer, she feels, strangely, safer than before. She says, "I keep thinking, well, lightning can't strike twice."

Bob Desjardins left Air Canada in 1985 to concentrate his energies upon Distribution Nadair, an import-export company he owns in partnership with a friend.

The incident brought flight attendant Annie Swift closer to key decisions about her future. She ended her relationship with her boyfriend, gave up her suburban house, and bought a country home in St. Lazare, near "Uncle" Ludwig's farm, where she could spend her free time horseback riding. She lived there for two years before she married Air Canada pilot Kent Russell in January 1988 and moved to Vancouver. She continues to work as a flight attendant for Air Canada.

"I felt cheated. . . ." recalls flight attendant Susan Jewett, ". . . cheated of my innocence of flying." She took a leave of absence from her job, to work out her anger at the people, procedures, and coincidences that had jeopardized her life. Gradually the anger was replaced by gratitude that her life had been saved by the skills of Captain Pearson and First Officer Quintal. She resumed work as an Air Canada flight attendant, but there is one key difference in how she performs her job. "If there is a mechanical delay, I want to know why," she says. She is now an in-charge flight attendant.

Jewett's second daughter, Elizabeth, was born in August 1985.

Flight attendant Danielle Riendeau took an eight-month leave of absence. She suffered from a deeply bruised shoulder, incurred when she had to crack open a reluctant emergency exit door. Her emotional scars ran deeper, however. When she returned to work she was in the unenviable position of being on reserve, meaning that she could not choose the particular flights—and aircraft—that she would fly. The first time she was assigned to a 767 she dissolved into tears. In an effort to help, Bob Desjardins, the in-charge flight attendant on Flight 143, arranged a dead-heading trip, on which he and Riendeau could ride as passengers on a 767. She cried throughout the flight, and worried that her career was over.

"I didn't want to lose my job," she says. "There is no other job that gives you the time off and the opportunity to meet people and see different places. I didn't want to go back to being a secretary."

She entered into therapy, learning behavioral techniques to deal with her fears. For example, she wore a rubber band on her wrist. In flight, when she was overcome with fear, she learned to snap the rubber band against her skin, hard. "It kind of wakes you up," she says.

Over time the therapy worked, but three years passed before the legacy of fear became truly manageable, and a residual effect still lingers. "I'll never be the same flight attendant," she says. "Never. But I've gone a long way. Now I don't think about the bad things, and the 767 is my favorite airplane. When I fly Aircraft 604 I have this little warm spot in my heart. That airplane saved my life."

Captain Bob Pearson and First Officer Maurice Quintal received the plaudits of their peers, both nationally and worldwide. Among the honors bestowed upon them were Certificates of Merit from the Canadian Air Line Pilots' Association and Outstanding Airmanship Awards from the Federation Aeronautique Internationale.

During another incident, however, they wondered whether they would have the chance to qualify for more awards. On June 23, 1984, eleven months to the day after the Gimli incident, Pearson and Quintal took off in the same airplane, Aircraft 604, from Ottawa to Montreal. During the initial climb away from the airport, the very same sequence of warning buzzers and lights commanded their attention. "Here we go again!" Pearson said. But when he leveled off in preparation for a quick return to Ottawa, the warnings ceased. Safely back on the ground, Pearson learned that false warnings had been set off as a result of the 767's steep rate of climb. "A few four-letter words were on *that* cockpit tape recorder," he recalls.

It is not true that the Canadian government named an airport after Captain Bob Pearson. In 1985 the name of Toronto International Airport was changed to Pearson International Airport, but that was in honor of former Prime Minister Lester B. Pearson, no relation to the pilot.

In fact, little has changed for Pearson. He is still a 767 pilot for Air Canada. He is completing work on the country home he is building and still loves to level an opposing hockey player with a well-timed body check. In his and Carol's future, he sees only blue skies.

He was, perhaps, less affected than the passengers because he—and to a somewhat lesser degree Quintal—was in control. He knew the scope of the problem. He did not know the outcome, but he knew he could influence it.

"Sometimes I think that I had the easiest job in the airplane," he says. "If I had been sitting in the back with the passengers, I'd probably still have the seat stuck to my pants."

One year after the Gimli incident, Maurice Quintal's wife died, and he faced the future as a single parent to his boys, Jean-François and Martin.

Quintal struck up an acquaintance with Claudette Plouffe, the executive director of the Commission of Inquiry that investigated the Gimli incident. After a lengthy courtship, they were married in 1986.

Quintal's career similarly bounced back. The handy excuse in any airplane incident is "pilot error," and Air Canada initially attempted to place partial blame upon both Pearson and Quintal for their role in the fueling process. But the full inquiry cleared them and, in fact, praised their airmanship. As a result, Quintal, in 1989, began the training that will allow him to assume the captain's seat.

Postscript: Who Was to Blame?

Who was responsible for sending Flight 143 on its way with only half the necessary fuel load? After an internal investigation, Air Canada divided the blame between Captain Pearson, First Officer Quintal, and mechanics Conrad Yaremko, Jean Ouellet, and Rodrigue Bourbeau. But when the airline announced a variety of suspensions and other disciplinary procedures, the pilots and mechanics fought back through their respective associations with the support of the Canadian press.

The ensuing outcry caused the government of Canada to establish a special independent Board of Inquiry. After hearing the testimony of 121 witnesses during sixty-five days of hearings, conducted intermittently between November 15, 1983, and November 8, 1984, the board issued a 199-page final report concluding that the incident was not the fault of the pilots or the mechanics, and cited various deficiencies in airline procedures. Among them were:

- the decision to introduce a new aircraft that weighed its

fuel in kilograms, while other aircraft in the fleet continued to weigh their fuel in imperial pounds. Air Canada made this decision at least in part under the urging of the Canadian government, which wished to advance the national metrication campaign. But the board's final report declared, ". . . if Air Canada had been at all concerned about flight safety, it would have resisted pressure from any direction, including the government of Canada . . . and would have retained gauges in imperial measurement."

• the failure to assign responsibility for calculation of fuel. Pilots testified at the hearings that they had been told that fuel calculation on the 767 was the responsibility of maintenance; for their part, maintenance personnel contended that they were told it was the pilots' duty. The report concluded, "Air Canada failed to fill the gap left by the departure of the second officer."

• the failure to train either cockpit or ground personnel to calculate the 767's fuel load in the event of the failure of the fuel quantity processor.

• confusion concerning the Minimum Equipment List. The board declared, "A comparison of Air Canada's MEL with those of other carriers which operate the 767, such as Delta Air Lines, TransWorld Airlines, United Airlines and American Airlines, shows at a glance that they have MELs which are much more clear and precise."

• the lack of spare parts. If mechanic Conrad Yaremko had been able to replace the faulty fuel quantity processor in Edmonton the night before Flight 143, the untoward chain of events would never have occurred.

In sum, proclaimed board chairman Mr. Justice George Lockwood, "The evidence of a failure of communications at all levels of Air Canada is alarming. While this may in fact be a problem with all large corporations, it is of particular concern in an industry which is daily responsible for untold numbers of human lives."

Lockwood went on to absolve individual pilots and mechanics from blame, and added, "The consequence would have been disastrous had it not been for the flying ability of Captain Pearson with valuable assistance from First Officer Quintal. Ironically, the avoidance of disaster was, to a considerable extent, due to Capt. Pearson's knowledge of gliding. He applied such knowledge to the successful flying and landing of one of the most sophisticated commercial aircraft ever built.

". . . thanks to the professionalism and skill of the flight crew and of the flight attendants, the corporate and equipment deficiencies were overcome and a major disaster averted."